Language of the Universe ... Maybe

Language of the Universe . . . Maybe

J.D. WATERS

AuthorHouse™ UK
1663 Liberty Drive
Bloomington, IN 47403 USA
www.authorhouse.co.uk
Phone: 0800.197.4150

© 2015 J.D. Waters. All rights reserved.

No part of this book may be reproduced, stored in a retrieval system, or transmitted by any means without the written permission of the author.

Published by AuthorHouse 04/14/2015

ISBN: 978-1-5049-4010-8 (sc)
ISBN: 978-1-5049-4009-2 (hc)
ISBN: 978-1-5049-4011-5 (e)

Print information available on the last page.

Any people depicted in stock imagery provided by Thinkstock are models, and such images are being used for illustrative purposes only.
Certain stock imagery © Thinkstock.

This book is printed on acid-free paper.

Because of the dynamic nature of the Internet, any web addresses or links contained in this book may have changed since publication and may no longer be valid. The views expressed in this work are solely those of the author and do not necessarily reflect the views of the publisher, and the publisher hereby disclaims any responsibility for them.

This book is dedicated to the Musicians of the World

Preface

This book contains a large number of scientific hypotheses, none of which have been subjected to empirical research or testing. As such, the book will probably be considered of doubtful scientific value by experts.

The normally accepted scientific tradition, founded by Sir Isaac Newton, involves the full process of evaluation through empirical testing followed by the publication of the results in a blaze of glory.

The concept of keeping important scientific hypotheses secret, as Sir Isaac Newton was fond of doing, is doubtful. If Newton had published his preliminary results regarding the orbit of the moon, other scientists of his day could have helped to formulate the gravitational theory. Of course, this might have meant that Newton would have missed the individual glorification which many scientists seem to aspire to.

If I were to follow Newton's example, it would take me a lifetime simply to test one or two hypotheses. The rest would then die with me. In our interdependent society I believe all scientists should work together as much as possible to test and refine those hypotheses. Only by combining our efforts can we achieve the evolution of knowledge which may hold the key to our survival. After all, who needs individual glory in an extinct society?

Chapter 1

Career Prediction

In May 1975, I was a Trainee Careers Officer studying for my Diploma of Vocational Guidance at the South Bank Polytechnic, London.

During that month I had been asked to compile a list of questions which would help me to interview school-leavers who needed vocational guidance. Up to that time, I had used my tutors approach. Now I had to think up my own approach.

First I had to decide upon a definition of vocational guidance. I decided to use the Employment and Training Act, 1973 as my main guide, and I paid particular reference to the definition of the vocational guidance service given to British citizens by the government information centres.

The Citizens' Advice Bureau, which informs parents and school-leavers of their rights in this respect under British law, stated in its handbook - "Young persons under 10, or over 15, and still at school may obtain advice on what is likely to be the most suitable career" (CANS - 1975).

Thus school-leavers would expect advice on "what is likely to be the most suitable career..." Careers were defined in my dictionary as "occupations leading to success". Fine, but what was meant by "most suitable?"

That question had been answered for me a month or so earlier by Miss Rachel Leyman, a Specialist Careers Officer at my employers' offices in Willesden, London. I had asked Rachel two questions, namely:

"What do school-leavers seek in a career?" to which Rachel had answered - "Status, variety, interest and jobs with people".

"What makes a good Careers Officer?" according to Rachel, the answer lay "somewhere within the realms of philosophy and psychology".

In compiling my interview questions, I could manage those aspects of a career concerning status, variety and jobs with people; but how do you define 'interest'? This was a problem. People are interested in so many different things. The question of why they are interested in one thing rather than another would need an explanation, -- if satisfactory interview questions were to be compiled. Presumably the answer lay in the realms of philosophy and psychology, as Rachel had said.

I decided to answer the question of Interest by going back to first principles. I looked up the dictionary definition of psychology - "a branch of science studying the processes, motives, reactions and nature of the human mind". All right. But what did science mean? I couldn't remember. The answer - "any branch of knowledge based on systematic observations of facts and seeking to formulate general explanatory laws and hypotheses that could be verified empirically". Great.

In discovering the answers to vocational needs, I was really involving myself in career prediction. I was satisfied that the answer to a career prediction hypothesis lay in the correct judgement of a school-leaver's ability, personality and job environment.

Ability could be tested fairly satisfactorily. The job environment could be easily determined, although not necessarily changed to suit the needs of the school-leaver. So that left personality.

I considered that personality and interest were essentially the same thing. Personality decided interest and interest decided personality.

The words "verified empirically", used in the definition of the word 'Science', indicated the effort which lay ahead in the verification of any hypothesis I might dream up. I would have to make a large number of observations of facts and amass a great deal of evidence. Bearing this in mind, I decided not to be too ambitious. I would look for something very simple and basic.

I reflected on the differences between myself and my brother. I have tended to be a bit of a day dreamer while he tended not to day-dream. I thought back to the time when I started to day-dream. I used to rock in my sleep from about the age of three onwards. My brother did not rock. I was awake as I rocked - my head resting on my fists - crouching under my eiderdown and rocking from side to side. I used to do this before going to sleep and before getting up in the mornings. I've seen other children doing the same thing.

So why was this? What caused it? Why didn't my brother rock or daydream? Did the day-dreaming make us different, and if so, how?

I worked out a hypothesis based upon day-dreaming, and tried it out on my fellow students. It was partly successful out not good enough to be right. So back I went to the drawing board.

Chapter 2

Performing Babies?

I decided to look at the way children were brought up in their very early years, and to look for simple differences. So what does a baby do in its early months after birth? According to books on the subject, it acts and reacts to its environment. If there is no human contact, it will die. Adult responses to a baby's needs are clearly very important.

Adult responses could provide a very basic and simple difference. All babies grow up under one of two conditions. Either they are brought up by a single parent, and respond to only one adult as they grow; or they are brought up by more than one parent, or adult, and consequently respond with more than one adult as they grow. A very simple & basic difference. Careful questioning might well reveal which situation existed for a school-leaver.

I visualised the two different situations. In the first case, the baby would cry and the single adult would respond with attention, food, nappy changing etc. As only the same single adult came each time, the baby would learn only to expect that one adult, when it cried, and thus would only learn to respond with one adult as it grew up.

In the second case, when the baby cried, one of several adults would attend to it. The adults might include the father, grandmother, elder sister or aunts, as well as the baby's mother. This baby would learn to expect a variety of adults, and would respond differently with each. It would also learn which cry brought the largest number of adults.

After some thought, I decided that all babies were natural performers. They acted and entertained while the adults applauded. Sometimes the responses were good and sometimes they were bad. By trial and error, the babies would expand their repertoire of attention seeking performances.

I classified the baby brought up by a single adult, as an individual response seeker, or I.R.S. for short. The baby brought up by more than one adult was classified as a public response seeker, or P.R.S. Individual performers or public performers. Thus all babies were either an I.R.S. or a P.R.S. and since babies grow up to become adults, this classification would apply to everybody.

I was well satisfied with this basic hypothesis, as I could recall different types of children who exactly fitted these definitions. There was the sort of child who stood in the middle of a room and banged a toy drum, looking round to make sure everybody had noticed. Then there was the other sort of child who tugged at your sleeve to show you something secret - something only the two of you would know about. One child required a large audience, and knew how to get their attention; the other required an audience of one.

This hypothesis left me with the problem of the day-dreamer type, or the non-response seeker, still to be explained. I thought back to my brother and I. What about my relatives, my uncles and aunts, my cousins and my friends? What were the obvious, simple, basic differences?

I am older than my brother, by two years. That is one basic difference between us. Would that be too basic? No, nothing could be too basic for my hypothesis. Older or younger - that would be a difference ...or rather elder, youngest or only child. That would apply to everybody. So what is the difference between the elder children and the others?

All children start off the same, in so far as they are either only children or the youngest child. They become elder children when another child is born or adopted. So what happens to an elder child in this situation? They are either an IRS or a PRS. They are used to performing and getting a response particularly from their mother. Then the next baby arrives, the mother must attend to that new baby. A mother cannot respond to both the elder child and the new baby at once. If the elder child demands first priority, it is likely to be rejected.

I decided that this was the vital difference between the elder children and the others. Depending on their age difference, all elder children would feel some element of rejection. The smaller the age difference, the greater the rejection feelings. IRS's would feel greater rejection than PRS's.

So what happened after rejection? The performer had performed as usual but the audience had not reacted. The elder child had demanded attention but the mother was too busy with the new baby to respond.

I concluded that the child performer would blame either itself or the audience (mother) for its failure to get its usual response. If the elder child blamed itself - it would try to understand where its 'act' had gone wrong, and it would try to improve its 'performance'. Alternatively, if the elder child blamed the audience for its failure it would try to cure the audience's "deficiency".

The children who tried to understand their failure and improve their performance, I called knowledge seekers. The Individual Response Seekers became Individual Knowledge Seekers, or IKS's; while the P.R.S.'s became Public Knowledge Seekers, or P.K.S.'s.

The children who tried to 'cure' their audience, I called cure seekers. Individual Cure Seekers and Public Cure Seekers. ICS's and PCS's.

The knowledge seekers would become the natural and social scientists of the world, while the Cure Seekers would become the psychopaths of the world.

Now I had a very simple hypothesis based on childhood responses to adults. Although it was very simple and basic, it had already presented me with six personality types to look for in my career prediction empirical research.

I decided to rule out the cure seekers from my vocational guidance research.

They would remain part of the hypothesis of course, but I hoped they would not be among the school-leavers I was to interview: This left me with four types, namely:-

1. IRS's - Individual Response Seekers
2. PRS's - Public Response Seekers
3. IKS's - Individual Knowledge Seekers
4. PKS's - Public Knowledge Seekers

I have listed the likely career predictions for these types In Appendix A at the back of the book.

Chapter 3

Blood out of a stone

Before putting my hypothesis to the test, I wanted to be sure in my own mind that it was more or less valid.

I remembered the old saying -- "You can't get blood out of a stone". This saying is taken from a fairy story where a little prince challenged a giant to a contest of strength. The one who could squeeze blood out of a stone would be the winner. On the night before the contest, the little prince collected some blood oranges (a type of orange whose juice is red in colour), and rolled them in the dust so that they looked like stones.

The next day, when the contest started, the giant picked up a large stone - and squeezed, and squeezed, and squeezed; but no blood came out. The little prince then picked up one of the dusty oranges -- and squeezed out the blood red juice. Thus the giant was defeated and the prince went on to greater things.

The reason I remembered the saying -- and the story, was that I considered that no personality types could exist today, unless they had existed in our ancestors. Thus you can't get blood out of a stone -- "unless there is blood already in the stone". Equally, there could not be response seekers today unless they existed in our historical past. History was easy to check -- thanks to the work done by our historians and archaeologists. If the six personality types existed in the past, then the Hypothesis would be on a much sounder footing.

Before starting to check through my history books, I decided to make a number of predictions by defining the personality types and considering the effect they should have historically. If the predictions proved to be correct, the hypothesis would be immensely strengthened.

The base types in any society would be the IRS's and PRS's. If the others were to emerge, there would have to be conditions existing within the society which would allow the elder children to feel rejected.

From 1968 to 1972, I had lived in Australia. I spent two years at a place called Groote Eylandt, in the Northern Territories. I had been able to observe a tribe of aborigines living on the island and had talked to some of them about the way their society was organised.

Under tribal law, everything is communally owned, except wives and some personal property like spears etc. The children are looked after by their mothers and the older women in a communal way. Everything is communal. The hunting the gathering of food, the cooking and mating, washing and dancing -- all is done communally.

In these communal societies, the children grow up among many adults, and since children are precious to the survival of the tribe, they are not allowed to be rejected or to feel rejected.

Because of this communal system of child rearing I would predict that these societies would produce public response seekers. No knowledge seekers would ever develop. Such communal tribes would never improve their methods of survival by their own inventions. They could only learn by responding with other tribes.

The lack of home-grown knowledge seekers would mean that technical and social advances would depend upon contact being established with other more advanced tribes. If no such contact was made, there would be no technical or social advance.

This prediction flew in the face of modern historical theories which said that technical and social advances depended upon the availability of surplus food due to a favourable climate and soil. Nevertheless, upon checking with anthropological studies and historical records, I found that the prediction was fulfilled completely. All communal societies, regardless of climate, good soil and surplus food, stayed in the stone-age unless they had regular contact with people of more advanced cultures.

In order for knowledge seekers or cure seekers to develop, the elder children must be allowed to feel rejected. This can only happen if the children are left alone. These conditions would only be met if the children of the tribe were separated from each other, and from adults also. Such conditions do not exist in communal societies. The conditions are only found in individualistic societies, where there is private property, and where homes and gardens are separated from each other.

So the next prediction was that knowledge seekers, leading to technical and social advances on the one hand; and cure seekers leading to mass murder and criminality on the other hand, would only be found in individualistic societies. In such societies, there would be private property and individualistic enterprise resulting from private ownership.

Because of the individualistic enterprise nature of the society, the market place would be found in the centre of the community. This would contrast with communal societies, where any market place would be found between two tribes but not within the community.

I then checked through the historical records of sixteen ancient civilisations, including those of Egypt, China, Rome, Athens, the Incas and the Mayans. All had private property when they started. All had centrally placed markets in their settlements. All made technical and social advances, often completely independently. All had their fair share of mass murderers and psychopaths.

It was also noticeable that while the system of private property and individualistic enterprise remained in existence, the communities made continuous technical and social advances. If the private property system was abolished and a return made to communal enterprise, the advances came to a malt and the civilisation disintegrated.

Thus my second prediction was confirmed sixteen times over. Knowledge seekers and cure seekers do emerge within individualistic societies, and they disappear if the societies return to a communal system. This is well illustrated in the cases of China, Egypt and Rome.

Up to the time of the Chou Dynasty in Ancient China, the Chinese civilisation made continuous technical and social advances. The Chou Dynasty converted the private farmers into serfs, and from that time onwards China was governed on a feudal system. The farmers no longer worked for themselves and they lived in communal villages rather than their own private small holdings and farms. After the Dynasty had been established, civilisation and its attendant technical and social advances came to a halt.

The Egyptians were also private farmers until one of their Pharaoh's converted them into slaves. The slaves were organised into communal societies and the 'civilisation' stopped. The historical record of this conversion is in the Holy Bible (Genesis 47, Paras. 13-22).

The Roman civilisation stopped in a similar nay. The private holdings of the farmer citizens were converted into large estates and their previous owners were either killed or turned into slaves. Only the outward expansion of the Roman Empire, and the establishment of private farms by its soldiers, kept the civilisation going. Almost invariably good soldiers, administrators and emperors came from the outlying parts of the empire rather than the centre. When the private system collapsed, the 'Civilisation' collapsed also.

As I have already indicated, the communal system produces mainly public response seekers. These are essentially public performers of one kind or another. They all like large audiences to perform to and respond with. Their occupations depend upon their confidence and ability, and the two are often inter linked. Public response seekers can be classified according to their prominence in front of an audience.

In the front rank are the actors, entertainers, dancers, orators, lecturers, priests and soldiers etc., that actually appear in public to do their 'act'.

In the second rank are those whose work is publicly seen and applauded, but who do not themselves emerge on the 'stage'. They include playwrights, authors, journalists, photographers, artists, craftsmen, architects, builders, farmers and manufacturers etc.

The individual response seekers are mainly gossips etc., who need a one to one situation and a certain amount of privacy. They may include secretaries, salesmen and shopkeepers.

The knowledge seekers would like an audience, either of individuals in the case of IKS's or the public In the case of PKS's. However, they get their security from learning about things, and by continually trying to improve themselves and their environment. The individual knowledge seekers may become good teachers or nurses etc. The public knowledge seekers will prefer to research into improvements in the environment -- which can be publicly acclaimed. They will probably aim for Nobel Prizes etc., and can be expected to publish their work.

The cure seekers aim to correct the deficiency of the individual or public audience which has not responded as the cure seeker thinks it should have done. All sorts of curative measures will be tried to ensure a consistent audience reaction. Experienced cure seekers can be expected to go for simple remedies which produce a predictable audience reaction. They are usually classified as 'mad' or 'deviant'.

I would have expected people with a background like that of Adolph Hitler or Joseph Stalin to be public cure seekers. Both grew up initially with two adoring parents and as such they would have been public response seekers as young children. Then the fathers of both started to react badly. So their mothers would continue to praise and applaud them while their fathers did the opposite. Naturally, since one half of the audience was still applauding the children could conclude that their 'acts' were okay.

Therefore they would blame the 'bad' half of their audience, and become public cure seekers aiming to correct the 'deficiency' of their audiences. (Their fathers have a lot to answer for). I expect Crippen and Haigh were individual cure seekers.

I made further predictions on the basis of population growth or reduction. In an individualistic society, a low birth rate, or a high infant mortality rate would tend to produce mainly PRS's or IRS's. This is because 'only' children or the youngest child are invariably of this type. A high birth rate, or low Infant mortality rate would tend to produce more knowledge seekers and cure seekers. This is because there would be more elder children.

Thus a rapid increase in population would tend to produce knowledge seekers etc., providing the society was an individualistic one. This prediction has also been confirmed historically. Such population increases are followed by periods of technical and social advances, and wars also.

So now I had made two predictions, based on my hypothesis, one concerning the relationship between technical and social advances, and individualistic societies; and one concerning the effects of a rapid population increase in such societies. Both predictions had been completely fulfilled.

Despite this, I still wasn't completely satisfied. I wanted to be on a really firm theoretical base. After all, the hypothesis suggested some substantial implications for our own society, and if I published the hypothesis at any stage, there might be unforeseen repercussions. I recalled the 'blood out of a stone' saying and thought of the need for an ancestral base. This pointed back to our pre-tribal ancestors which were supposed to be apes of some kind. Did apes also have response seekers? If my hypothesis was right they should have done..........

Chapter 4

Apes and Birds

I considered the apes I had read about or seen films of. It was clear that apes were individual response seekers. Ape babies are brought up by their mothers only. There is no 'pairing' among the adults, except sometimes around mating time. The baby ape grows up responding only to its mother, because it takes about five years for the baby to be weaned-off its mother's milk, and on to other foods. In this time, the mother is the baby's sole provider of food as well as its main source of love, play, protection and transport. Inevitably, the baby ape grows up to be an individual response seeker.

All apes tend to respond to each other on a 'one to one' basis. They have a 'pecking' order which is determined by individual confrontation. The males and females are 'ranked' in this way. Mating is achieved after an individual display charge. Several males may mate the same female, but everything is done according to rank on a series basis.

While I was thinking of apes and their rearing patterns, two unrelated thoughts flashed into my mind. One concerned birds, which appeared to be public response seekers. The other thought concerned the major difference between humans and all other animals regarding their rearing patterns.

Most birds 'pair off' for mating, and the subsequent rearing of their young. This would suggest that young birds would grow up to respond with more than one adult. Such birds would become public response seekers and

Language of the Universe . . . Maybe

therefore public performers. I could imagine that their courtship displays and territorial songs might be examples of such public performances. However, many of the mating dances etc. are not learned. This suggested that the behavioural responses of the birds might also include a structural element. Could such a structural element be found in apes?

I then considered the pattern of response seeking. Birds were PRS's. Apes were IRS's. Communal societies of humans were PRS's, while individualistic societies of humans were essentially IRS's and PRS's. There seemed to be a certain pattern.

While thinking of patterns, I turned my attention to the other thought which had crossed my mind. This thought concerned human rearing patterns -- and the fact that I had not thought of it before will give the reader some idea of my stupidity.

I had been looking for simple differences between the IRS apes and the PRS humans (of the communal society), -- looking for simple differences in behaviour; and I had not noticed the most obvious, basic, simple difference. It is a difference which makes humans distinguishable from all other animals. It is not only physiology -- it is behaviour.

All mammals except humans rear their young to semi-maturity before having another brood. Humans by contrast, may give birth to child after child long before the first baby is semi-mature. Apes have their young at five year intervals, while humans may have one child every nine months.

This is a very significant difference from a behavioural response point of view - because it means that only humans will have 'rejected' elder children and therefore knowledge seekers. All other mammals will remain IRS's or PRS's, if they have response characteristics. Yes, a very simple and basic difference, but one that took a couple of hours to get through my thick skull. How stupid can you get?

This simple difference prompted me to consider the effect of a multi-age brood on a ape society. If humans developed from apes, then there must have been a behavioural change from the IRS apes to the PRS response. PRS's develop when babies are brought up by more than one adult. What could induce ape mothers to allow other apes to bring up their babies?

I decided the answer was directly connected with mating. If a female ape was born with a genetic defect in her 'oestrous' mechanism, so that she failed to come 'off' oestrous after mating, this could cause an 'apparent' pairing situation and also lead to a multi-age family of young apes. Then another thought came to me, how do apes stop inbreeding?

Chapter 5

Inbreeding

Human societies, whether communal or individualistic, have elaborate laws or tribal arrangements to stop inbreeding. As far as I know, other animals do not have such codes of law. How then, do they stop inbreeding? In the case of birds or insects, there were so many of them that they could mate on a chance basis and avoid inbreeding. But a troop of apes would tend to be much more parochial. Thus there would have to be a good deal of cross-breeding to stop inbreeding in an ape society. I hypothesized a possible solution.

It would be necessary for the larger and stronger males to mate the smaller and weaker females; and for the weaker males to mate the stronger females.

I decided that the high ranking males in a ape society would tend to mate the most timid females first. This would happen because the timid females would be the first to take up a sexual submission pose when the high ranking males made their display charges. As a result, the timid females would mother the sons and daughters of the high ranking males.

From a behavioural point of view, the timid, anxious female would be very responsive and therefore a very good mother. Her sons would grow up to be strong and secure, and as such would develop into high ranking males. Her daughters, however, would also be strong and secure, and as such would not be timid. Thus the timid females' daughters would not be mated by the high ranking males, because the daughters would be very slow to take up the sexual submission pose.

The main difficulty with this hypothesis concerned the mating of the strong, secure females by the low ranking males. After all, if high ranking males could not mate such females, what chance would low ranking males have? Despite this, I decided that the hypothesis was right and that somehow the strong females were mated by the low ranking, weaker males.

A strong female would be a bad mother. Being very secure herself, she would not be anxious and would not respond quickly to her baby's needs. Her offspring would grow up anxious and insecure. Her sons would become low ranking males due to their timidity, while her daughters would become the timid females which were mated by the high ranking males. Thus the cycle would be completed with a continual interchange of characters, thus preventing inbreeding.

Apart from the difficulty of the mating of the strong females by the low ranking males, the hypothesis would work very well from a behavioural point of view. I was very fortunate in that there had been some excellent field-work done on apes. Jane van Lawick Goodhall had researched into the characteristics of chimpanzees in the Gorme River area of Tanganyika (Tanzania).

In her excellent book -- "In the Shadow of Man", she has described the mating habits of these apes, enabling my hypothesis to be tested. An examination of the reality as presented in her book, shows that the hypothesis is broadly correct. My difficulties over the mating of the strong females is solved in an entirely logical manner, which emphasizes again my own stupid, illogical thinking.

What actually happens is as follows: --

The highest ranking males assert their authority by their charging displays. When such a male charges, the more timid apes run out of the way. The courting charge of the male is the same as the assertive charge.

When a female comes into oestrous, her bottom turns bright red and becomes quite swollen. As soon as the males spot such a female, they charge over towards her, with the highest ranking male to the fore. The more timid a female, the more quickly she runs out of the way. As she does so, she shows her bottom and this simply encourages the males to charge after her. The female runs away, simply to get out of the way of what appears to her to be a normal act of male assertiveness. But wherever she goes she is pursued by the high ranking males. Eventually she is caught, and a sort of gang rape takes place.

The highest ranking male mates her first and then allows other high ranking males to mate her. The low ranking males are chased away by the high ranking males. In this way the timid female is only mated by the high ranking males.

In Miss Goodhall's book "In the Shadow of Man", I would surmise that 'Flo' was just such a timid and anxious female. Because of this, Flo is a very attentive and responsive mother. By virtue of her favoured position among the high ranking males, Flo then became a high ranking female and dominated the other females. Flo's sons, Faben and Figan, became high ranking males as predicted. Flo's daughter Fifi, grew up to be a strong, secure female.

However, when Fifi came into oestrous, she became sexually aggressive. She flirted with all the high ranking males, positively thrusting her pink bottom at them. But the high ranking males did not immediately mate her. Instead they made calming gestures to show her there was no need to take up a sexual submission pose -- because they had not charged at her. If the high ranking males do mate such aggressive females, it is usually in a rather half-hearted manner. It seems they are not interested in aggressive flirts, but prefer the excitement of the chase -- provided by the timid females.

Fifi then turned her attention to the low ranking males. As the low ranking males were kept away from the timid females by the high ranking males, they eagerly mated the sexually aggressive Fifi.

Thus the hypothesis was partly confirmed.

The Gorme Stream studies have not been going long enough to determine whether the daughters of Fifi will become timid females and thus responsive mothers like Flo. I would surmise however, that 'Marina' was a sexually aggressive female like Fifi. Marina was not a good mother. Her sons and daughters grew up to be timid and anxious. Marina's daughter was called 'Miff'. If my hypothesis was correct, I would have expected Miff to be a good mother. This prediction was in fact confirmed, although Miss Goodhall's team were very surprised at Miff's good mothering qualities. It will be interesting to see what sort of mother Fifi turns out to be.

I had now solved the problem of inbreeding to my own satisfaction and I went on to consider the effect of a female whose 'oestrous' mechanism would not switch off, because of a genetic defect.

Chapter 6

'Defective' Females?

Normal female apes come into oestrous approximately once every five years. During the interval they continue to produce milk and rear their babies to semi-maturity. After five years, they come into oestrous again, are re-mated and give birth to another baby.

A defective female, who did not come out of oestrous after mating, would continue to be mated up to and after the birth of her baby. In the following five years, when normal females would have had no more babies, the defective female would have one every nine months or so. This would lead to the following repercussions: --

1. The daughter of the defective females would probably inherit her defect.
2. The babies of the defective female would grow up surrounded by a lot of males because of their continuous interest in such a female. This would create an 'apparent' pairing situation, increasing the security of the babies. Thus the sons of such females would become high ranking males, while the daughters would become sexually aggressive females.
3. The sexually aggressive daughters of the defective female, and their daughters would mate for generations with the low ranking males, thus reducing the dimorphism of the apes.
4. Under the influence of the 'apparent' pairing effect the babies of such females would become public response seekers.

5. The much larger family produced by the defective female would require more food or higher calorie foods. More food would mean a much larger area would have to be covered in the ape society to obtain food supplies.

Higher calorie foods would indicate a change from a mainly vegetarian diet to a mainly meat diet. Both the hunting and food gathering activities would encourage the evolution of apes with a high power to weight ratio, and as such there would be a premium on good ground level locomotion.

Chapter 7

Individualistic Societies

To help me with the problem of the transition of a communal society into an individualistic society, I turned to Jacob Bronowski's book "The Ascent of Man". I did not see the television series as I do not have a television, but my mother had given me a copy of the book at Christmas.

Although I disagreed with many of Bronowski's conclusions, I found the book useful for reference purposes. Bronowski's facts may be right even if his conclusions are wrong.

In Chapter 2 of "The Ascent of Man", Bronowski's describes the life of the Baktiari nomads and concludes that civilisation 'can never grow up on the move'. I disagree. I would predict that individualistic societies would emerge from just such nomads as the Baktiari.

In my view the survival methods adopted by communal tribal societies would depend upon their environment.

In equatorial forests, like the Amazon basin, where game and plant food are plentiful, I would expect hunting and food gathering to be the basis of life.

In sub-tropical regions, like the savannah of Africa, there are less trees, and grass can grow in abundance. This encourages grazing animals, and the society would tend to consist of hunters and herders.

In temperate forests, there is little grass, and the society would tend to be of the plot clearer and planter type.

In deserts, the limited grazing would mean that the tribal societies would have to herd the animals, in order to reduce the otherwise enormous hunting distances involved. Herders are essentially hunters with a captive herd of game.

Communal tribes in poor desert regions would be forced to split into separate families in order to ensure that each of the society's herds had sufficient grazing grounds to survive. I would make the point that a communal society of herders would have several small herds rather than one large one. This is because the male animals, such a bulls or rams, cannot coexist with other males. Two bulls or rams in the same field will fight. Nomads would therefore be forced either to split the bulls and rams, or to castrate them. However, to rely on one bull or ram and a single large herd would be suicidal. If the bull or ram died, before it could mate the females, the whole herd would die out and the tribal society would die with it. Thus the herds are broken up into small units, with one bull or ram to each herd or flock.

So desert tribes would be forced to separate in order to ensure the tribes collective survival. Each herd would tend to radiate from a water hole which would serve as a family base. The tribes would probably meet once a year to make up for any losses in rams or bulls, and to conduct normal communal activities.

The children of such separated families could grow up separated from other children in the tribe, and thus elder children could feel rejected. Therefore, only in desert nomads, like the Baktiari, would I expect to find the emergence of knowledge seekers and cure seekers. Civilisation would develop on the move. There would be plenty of time for knowledge seekers to plan improvements while guarding their flocks by night.

As the people of these tribes would almost certainly supplement their diet with any high calorie foods they could find in the desert, it would not take many generations of knowledge seekers to create an agricultural civilisation -- given the opportunity. Only knowledge seekers would take the opportunity.

We know that the desert tribes of Israel had to split up into families because of poor grazing (Genesis Chapter 36 Para. 7). We also know that these tribes knew all about cross breeding of animals (Genesis Chapter 30). I think we can assume that the cross breeding of desert grasses to produce wheat was also the product of desert tribes. A survival conscious people like desert nomads would have the motive to grow such wheat around their water holes.

I regard it as highly significant that all sixteen civilisations I checked, had developed close to desert regions. They all retained herding animals as part of their agricultural system. All were animal herders turned planters. Chinese, Egyptian, Inca. All had different high calorie crops. All appeared to have developed separately. All had herding animals. All used irrigation extensively, suggesting a background consciousness of water conservation.

Thus the behavioural link was completed. In ape societies there were individual response seekers. In communal societies there were public response seekers; and in individualistic societies there were the six types found in 'advanced' western civilisations.

I had been brought up to think that only primates, like humans for example, reasoned things out. All other animals were instinctive and simply reacted in a mechanical sort of way to circumstances. Primates were supposed to be different from birds, reptiles, fish and insects. But were they so different? Birds had the same behavioural characteristics as the humans of communal tribes. Some fish are territorial and fight on a one-to-one basis. Could there be some structural elements, which were common to all animals, including humans, which our society had tended to overlook in its apparent desire to elevate itself on to a higher plane?

I pondered over the 'blood out of a stone' story, and thought of public response seeker birds. Was there a structural link?

Chapter 8

Structural Influences

The possibility of instinctive behaviour made me wonder about the correctness of my hypothesis.

There is a lot of controversy among psychologists about whether people behave mainly through instinct or by learned responses. My hypothesis had been based solely on responses, -- between a baby and its parents; and between subsequent adults in society. However, the pattern which had been shown to exist suggested structural influences affecting the evolution of behaviour.

At this point I decided to review the position and consider the philosophical aspects of my hypothesising, before continuing my research.

The 'blood out of a stone' story had forced me to consider of origin of personality types. The historical search had resulted in the confirmation of my hypothesis and revealed some interesting patterns in human society.

If civilisation depended upon the continuing emergence of knowledge seekers, and if this in turn depends on the right conditions for the development of such people, the lessons for our present day societies could be profound. Much of the mystery over the causes of the rise and fall of civilisations would disappear. Furthermore our attitudes concerning the apparent lack of intelligence of communal tribes would be corrected.

All of this information could result in a more understanding world. Wars and worry, caused by public response seekers in their roles as soldiers, priests and politicians, could be eliminated by a better informed public. On the other hand we might witness the development of some sort of behavioural engineering within our societies. Any new knowledge could be used for better or for worse.

This then, was a matter for my personal philosophy. I have always had a lot of faith in the common sense of ordinary people throughout the world; and while this newly discovered knowledge could be misused, I believed that most people would prefer to have the knowledge than remain in ignorance. The lack of knowledge in this area was already causing much distress, hardship and misunderstanding. Anyway, it was not my job to live people's lives for them or play 'god' with new information.

The question of structural influences was also a matter for philosophy. In my view we live in an evolving world. The environment evolves and we evolve as part of that environment. Our knowledge is continually developing and many beneficial advances have been made due to that evolution. It was clear to me that knowledge would not stop evolving simply because I chose not to think about it.

The 'blood out of a stone' story had forced me back into the past to primitive tribal societies, then to apes, then birds, then fish, then insects and finally to the cells of which everything is made. All characteristics must come from somewhere and must be transmitted in some way. This pointed to mating and reproduction.

In view of the importance of mating, I decided to determine the mechanisms of mating at its simplest. I felt this would show me whether there were any parallels between the structural mechanisms of animals and their behaviour due to responses.

In determining my career prediction hypothesis, I had reasoned things out myself first, and then checked history books etc., to see if my predictions would come true. They did, but at the cost of implying that existing historical theories about the development of civilisations were wrong. My hypothesis fitted the facts completely; the historians' theories did not. The question arose therefore, as to whether I should continue to hypothesis about the structural side of things, and then check the facts afterwards; or should I use the normal method of trying to deduce a hypothesis from an examination of all known facts.

Well, success breeds success, and I decided to stick to the dictionary method which had brought me success before. I would hypothesise first and check the known facts afterwards. After all, I might turn up something original.

I was spurred on by an idea which had come to me while I was considering the mating habits of animals. I had studied apes, birds, fish, insects and even multi-cellular plants, and found that they all shared a common characteristic of structural significance. They all mated in a longitudinal alignment.

I couldn't remember anybody pointing that out before. Perhaps it was too basic and simple to be worthy of attention. The obvious is always very easy to overlook, unless you happen to be very simple and stupid like me. The question or 'why' they all mated in-line intrigued me. I decided it was due to the possibility that the 'germ' cells mated in-line. Thus it might be found that the pollen and the ovule, or the sperm and egg join in-line.

The plants or animals developed, after all, from such germ cells, and as such could be said to be the children of such cells. Any characteristics of a basic kind are invariably passed on by parents to their children, and mating was a very basic characteristic. This seemed to me to be a perfectly logical piece of reasoning.

Thus I was confronted with the problem of working out the mechanism of germ cell joins. Before I started, I considered the difference between uni-celled organisms and multi-celled ones. Uni-celled organisms do not mate in-line.

In fact, they do not mate. They reproduce by splitting into two parts. I wondered whether they split in-line. I decided to work out a possible split mechanism. Then I would work out the cause of the split and finally decide how the cell joining would operate.

My eventual solutions to the above problems, led to the making or predictions in the usual way. When I checked these predictions against known facts, I found they matched almost perfectly. By using the same simple reasoning, I put together a multi-hypothesis and each stage so precisely interlocked with known facts that I decided to write this book.

The hypotheses are fantastic. They are very simple and very basic, and they could have profound implications - for better or for worse.

If anybody is bold enough to publish this book, the reader will be able to decide whether my decision to write the book was correct.

I'm afraid I've never been very good at writing, my grammar is poor and I seem to have a natural aversion to prose and poetry. As a Chartered Surveyor, my report writing invariably let me down. I disliked writing essays at school and I must say I dislike having to write this book. I apologise, therefore, for my poor style and presentation. I trust the reader will forgive me my poor use of the English Language and will concentrate on the subject matter herein.

Chapter 9

Universe and Nature

In this chapter, I explain my philosophical views on the relationship of the universe and nature as I see it. I regard this as necessary background to explain my approach to the structural problems involved in the hypothesis. In the next chapter, I discuss the influence of time, as I see it, because this is a structural dimension which is easy to overlook when making 'still life' drawings. I start my hypothesis on uni-cell splits, otherwise known as binary fission in Chapter 11.

In my opinion we live in the universe of Hydrogen. I believe that hydrogen is the basic element from which all the others are derived. Nobody knows the shape of hydrogen but I conceive of it as a tetrahedron, that is, an equilateral triangle in three dimensions.

The question arises as to where hydrogen came from. In my view it is a development of nature. This means I see nature as something different from the universe. I conceive of nature as being something which is random and unpredictable. It is so unpredictable that you cannot even rely on its unpredictability. Nature is composed of random and unpredictable times and dimensions.

Hydrogen, as I have said, developed from nature and it is therefore built of nature. Unlike nature, however, hydrogen builds in a regular, predictable way. Thus most things in the universe of hydrogen are likely to be predictable. However, because the building material is nature, the unexpected may happen at any time.

Language of the Universe . . . Maybe

Nature has dimensions which are finite and infinite. It also has dimensions which are finite at one end and infinite at the other. As hydrogen builds out of nature, these seemingly impossible and contradictory dimensions are found in the universe. In this way, most people can see only one end of a line at any one time. They cannot see both ends at once any more than they can see both sides of the moon. They know the moon is finite, but they might have a hard job proving it.

When I think of nature, I think of irony and paradox. I also think of the unexpected. It might be some humour which makes me laugh involuntarily, or it might be the sudden shock of an accident.

When I think of the relationship of nature to the Universe of hydrogen, I keep being reminded of Trinity College, Cambridge. I lived most of my early life in Cambridge, so I know the City fairly well.

Trinity College represents to me, the College of Hydrogen. Trinity means a group of three acting as one, and considering that my view of hydrogen is that it is basically triangular, the two seem synonymous.

When you enter Trinity College from Trinity Street, you will pass through the 'great' gate. Above the portals is a status of King Henry the Eighth, who founded the College. The great gate is an orderly, well-built octagonal structure, which reflects the precise regularity of hydrogen. The statue of King Henry shows him holding a stone orb in one hand and a wooden chair leg in the other. For most visitors, the chair leg is unexpected and causes a laugh. Nature strikes its first blow.

Having passed through the great gate, you will find yourself in a courtyard. It seems at first sight to be to be square, orderly, precise and predictable. There is an octagonal fountain in the centre. The paths are laid out in the form of a square, and are themselves formed of even, square paving stones. But then you notice, upon looking round, that the architectural styles of the buildings surrounding the courtyard are all different. Nature strikes again. You will further notice that the paths are edged with cobblestones, all random and unpredictable. Nature. However, to most people, the first courtyard seems somehow friendly and is very popular.

Then you walk through to the next courtyard. The architectural style is the same all through. The courtyard seems rather sterile and empty. The same can be said of the cloisters of Trinity College Library. This is considered to be an architectural masterpiece by some, but it seems rather cold and draughty. Most people hurry through to the 'backs'. The 'backs' is the name given to the River Cam as it winds its unpredictable way through the backs of the colleges. A nice, friendly river so it seems.

So this is the universe as I see it. The structure and regularity provided by the master builder hydrogen. The random unpredictability caused by the building material -- nature.

I have used the phrase "the universe of hydrogen" a number of times, implying that there may be other universes. So there may. However, it is doubtful whether we who are made from hydrogen, or its derivatives, could see beyond the confines of its structure. But with nature - you never know. Maybe.

I added the word 'maybe', because in my previous sentence I used the words "never know". It is a tendency of those who are built of the definite, structured materials of hydrogen to make definite, structured statements. Yes, we all make them. There – I've made another one.

With nature you know where you are and you never know where you are at the same time…. or at different times…maybe. It's very difficult not to be definite even when you are being indefinite. I trust the reader gets the picture.

Chapter 10

Time Projections, Time Lag and Time Relativity

Before I go on to discuss binary fission, I am going to consider some time projections. Time may be considered as a dimension which is linked directly to the other three. It is easily forgotten when structural drawings are made, and I will consider it first, so that we can have its influence in mind when considering binary fission mechanisms.

Time projections show the possible alternative places an object could have moved to, in any particular time lapse. Thus a man standing on the centre spot of the Wembley Football Stadium could walk ten paces forward in say ten seconds. In the same time, travelling at the same speed he could have gone in many different directions. A time projection shows where he could have got to, under these conditions.

Or course, the projections are in three dimensions while this book has only two dimensions. The reader will have to consider the presentations as sectional forms or three dimensional solids. The projections are illustrated in Figure 1 as follows:-

Figure 1   (a)   This represents a single particle moving in a straight line. It projects a sphere, seen here as a circle.

                (b)   This represents a single particle mowing in a curve. This projects a spiral.

(c) This represents a single particle mowing in a spiral. This projects a triangle, square, diamond, in fact all known crystal shapes.

(d) This represents a single particle oscillating backwards and forwards in a straight line. This projects a disc at right angles to the direction of the oscillation.

(e) This represents two particles circling a common centre point, in the same direction. This projects a two armed spiral.

(f) This represents two particles circling a common centre point, in opposite directions, one within the other. This projects an Ellipse.

(g) This represents two particles spiralling around a common centre point in opposite directions, one within the other. This projects a wave.

This last projection can be seen on Earth, in the action of rivers and currents. The river water represents particles spiralling inwards towards the centre of the Earth. If both the Earth and the rivers are considered to be moving, then each travels (relatively) in the opposite direction. Therefore rivers take the form of a wave.

In theory this should hold true for planetary paths also. (See Chapter 13).

Figure 1.

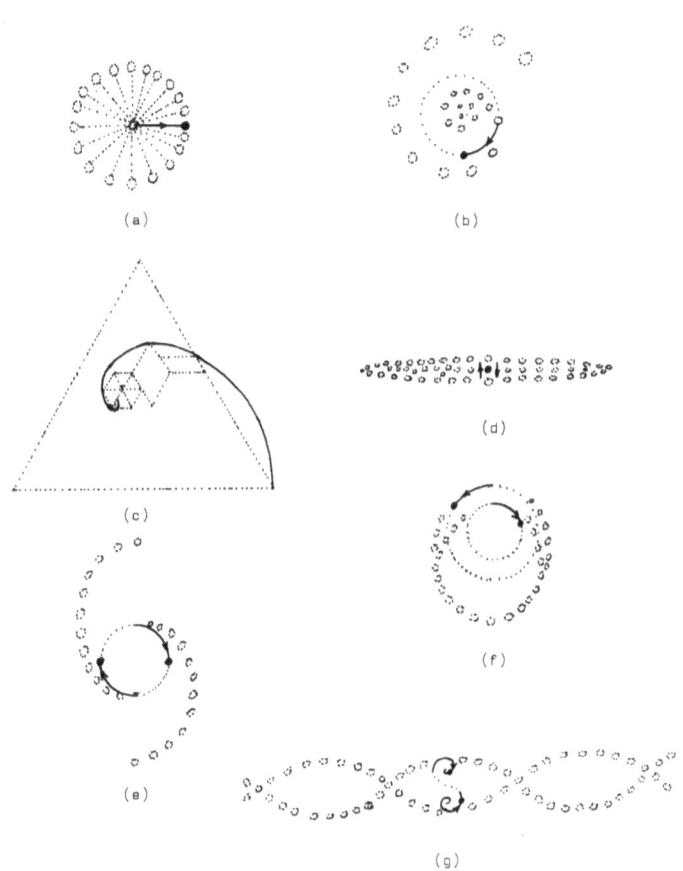

While I'm on the subject of time projections, I'll consider the evolution of some basic universal concepts. Thus we have Figure 2.

Figure 2

2(a) The circle represents the median between external and internal infinity.

2(b) The straight line represents relativity. It expresses the notional link between the external and internal infinities

2(c) The 'Y' represents divergence/convergence. Diverging from internal infinity to external infinity and converging from the external to the internal.

2(d) The 'H' represents a non-relative state, - that is a state without divergence or convergence.

2(e) The 'W' represents a state of continual divergence/convergence.

2(f) The circle with a dot in the middle represents two particles moving relative to each other without any divergence or convergence.

2(g) The equilateral spiral represents two particles moving relative to each other, either diverging or converging at a constant angle.

2(h) The spiral/straight line represents two particles moving relative to each other, either diverging or converging at an increasing or decreasing angle.

2(i) The heart-shaped figure represents two particles moving relative to each other, constantly diverging/converging.

So much for time projections.

Language of the Universe ... Maybe

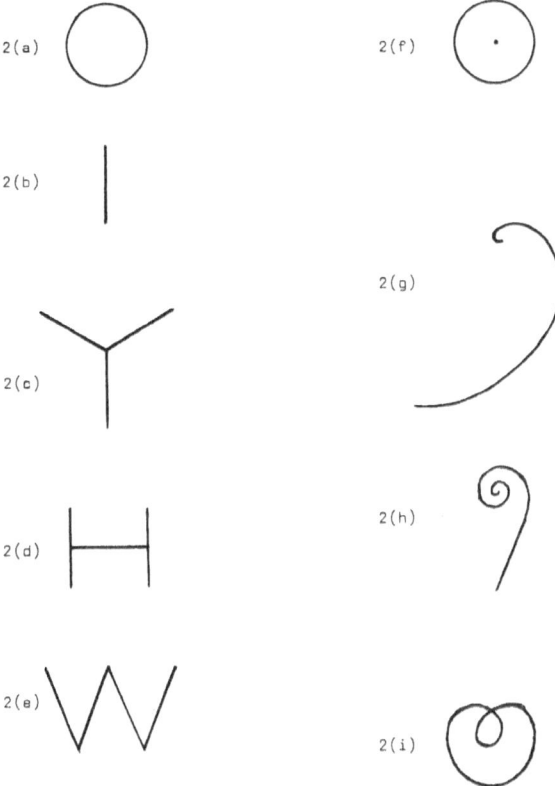

Next I would like to consider the effects of time lag. This is only important in relative terms. On a large scale it shows up in space. When we view galaxies through our telescopes, we see them as certain overall shapes. Spirals, ellipses and spheres are common.

However, we see these galaxies through the agency of light. Now although light travels very quickly by human standards, the enormous size of the galaxies means that the light reaching the Earth is of varying age.

If the galaxies were stationary, the varying age of the light would not matter. But if the galaxies are moving - and most Astronomers accept that they are - then the age variance becomes significant

Thus a photograph which shows a galaxy to be disc-like in cross section, as most spiral galaxies are thought to be, may present a distorted image due to the effects of time lag. Such galaxies are more probably cone-shaped, like a Chinese coolie's hat, with the apex of the cone pointing in the direction of overall movement.

I suspect the forces which shape galaxies are the same basic forces which shape the fruits and flowers of plants. Thus the spiral galaxies are like flowers, and the spherical galaxies are like fruits. The same basic forces produce the same basic shapes. If this analogy is correct, it suggests that the spherical galaxies are expanding while the spiral galaxies are contracting.

Finally I will consider the effects or what I call time relativity. For my purposes, this concerns the relationship of actual time taken as against possible time taken for a particle to travel from its starting point at any given speed.

Thus a particle moving in a straight line might move from point A to point B in one second. The time taken to get from point B to point A is also one second at the same speed. However, a particle travelling in a circle would take about three times longer to travel around the circumference as it would to travel across the diameter.

I express time relativity as a ratio or possible time to actual time assuming a constant speed. Thus the time relativity ratio for a straight line is 1:1. A circle is 1:1.507. A square is 1:6.74, while an equilateral triangle is 1:7.5.

In my view, it is time relativity which produces the oscillation affect and thus relative movement. Thus time relativity produces waves and differentiation.

Chapter 11

Binary Fission

The first thing I did regarding binary fission was to work out a possible structural mechanism by which it could work. I saw the uni-cell as a sort of house, with the chromosomes and nucleus as the inner furniture, and the cytoplasm etc., as the garden. I conceived of the garden and furniture built out of elements and compounds which had a positive or negative charge, and which were attracted to opposite charges on the inside or outside of the house. If the house were divided, the garden and furniture would be similarly divided.

For convenience I used a cube as a model to work on. I started by simply dividing the cube into two, as though I was slicing a wedding cake into two equal parts. This left me with two half cubes. Very simple and very basic - but not right. It was not correct because uni-cells split into two parts each of which is identical in shape to the original cell. Half cubes are not the same as cubes.

I had to think of a structure which could be split into two shapes each of which was the same as the original. Not so easy. After some thought, I decided that it would have to be a cube within a cube. Thus we have the basic cube in Figure 3(a). The X's mark the outside corners or the outer cube. Both the inner and outer cube have the same volume.

By reversing the outer cube, I ended up with two identical cubes, shape-wise - as in Figure 3(b). However, although they are identical in shape, they are not identical with the original cube in their composition. The former outside corners, marked by the X's, are now on the inside.

So we have a second reversal. As both cubes are now in the open, both can go through the second reversal process, as in Figure 3(c). Now we have two cubes each of which is identical with the parent cell in every way.

The chromosome 'furniture' would be attracted to the inside of the new uni-cells while the cytoplasm 'garden' would join the outside. Each cube has a hole in the middle which allows for the replication of the chromosomes etc., and each cell also has an external area which is the same as the original, allowing the cytoplasm etc., to double up in the same way.

This structural model seems to conform to the basic mechanisms of binary fission and would result in an even division of both the nucleus and the cytoplasm.

Unfortunately it does not work out in practice from a structural point of view. A cube does not, in fact, reverse into another cube. It works in two dimensions, in this book, -- but not in three dimensions. Nevertheless, the principle appears valid and the structure will work in three dimensions on a tetrahedron base. Thus we have Figure 3(d).

In some ways this may seem a rather complicated mechanism. It is in fact the simplest way to divide a structure into two parts in such a way as to produce an exact replica of the original structure.

As far as I could see from my researches, there has been no field work done on the structural mechanism of binary fission. The process has been observed and photographed but no three dimensional working models have been made which reproduce the process's structure.

However, by observing plants and animals, we can see the mechanism employed, particularly in reproduction and growth. In plants, this is the normal method of growth with the outer skin continually reversing to expose more inner skin. Leaves, petals, sepals etc., all reverse - turning their inside surface outwards. Many fruits split by reversal to reveal their seeds, and the seeds later split and reverse to allow the shoot and root to develop. Animals use the mechanism in mating and birth. Continuous reversal is a much used mechanism.

Language of the Universe . . . Maybe

Figure 3

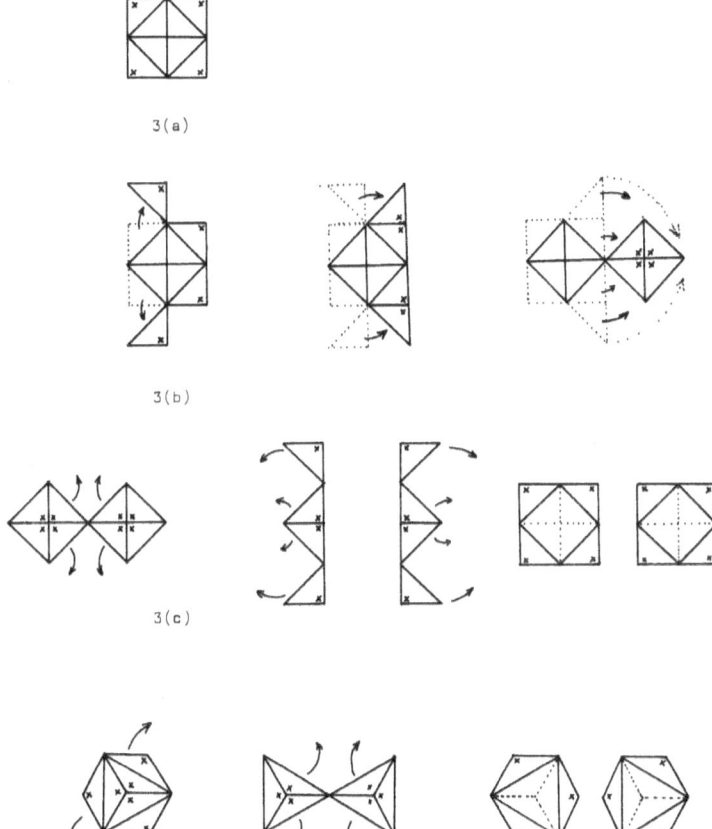

3(a)

3(b)

3(c)

3(d)

Chapter 12

Energy Spirals

In order to determine the cause of binary fission, I decided to consider the matter from a molecular point of view. This took me back to no-man's land again, as very little is known about the molecular or atomic bonding of structures. There are many theories, some of which closely fit known facts, but these do not have life models against which they can he compared. In particular, there are virtually no theories on the structural form of energy.

I hypothesised that energy took the form of an oscillating spiral, continually turning first one way and then reversing to turn the other. I have represented this as an 'S' shape in Figure 4(a). Perhaps the "Arabs" representation is more elegant. This is shown in Figure 4(b).

The energy spiral can he imagined as similar to a balance wheel spring in a clock, alternately contracting and expanding. This cyclical action is to be found at the heart or most things in the universe - indeed our own hearts are a good example of' such action. It is the same with our lungs and, of course, our cells.

As I have stated before in Chapter 10, the time projection of such motion takes the form of a wave. It is interesting to note that known examples of physical energy, such as photons and electrons, take the form of waves as well as particles. It is perhaps also interesting to note the 'heart' shape of the Arab number Five, which also represents such action.

Figure 4(c) shows that a particle which is continually spinning in one direction could follow an oscillating spiral route, without a reversal of spin.

While I was considering the atomic forces involved in binary fission, it occurred to me that the solar system might provide a good guide for explaining the mechanism of the split process. I therefore turned my attention briefly to the mechanism of planetary motion.

Figure 4

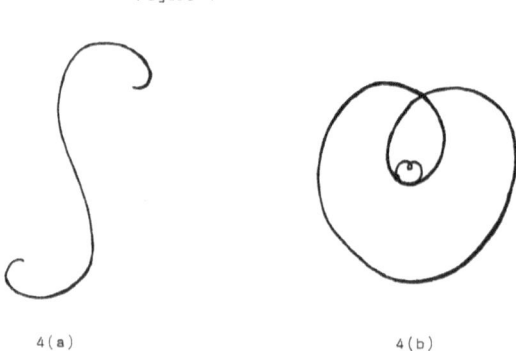

4(a)   4(b)

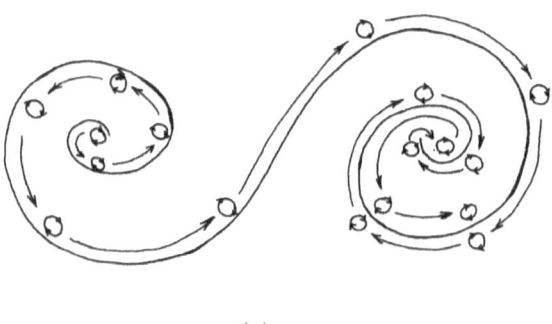

4(c)

Chapter 13

Planetary Motion

When we talk of atomic structures, we tend to think in terms of orbits and cycles. This is because the Astronomers' view of the mechanism of the solar system is used as a basis for atomic models.

The conception of the solar system, in terms of orbits and cycles, is fine for the purpose of Astronomy but it is a bit misleading in a true structural sense.

The Astronomers tell us that the moon revolves round the Earth, and the Earth revolves round the Sun, and that both the moon and Earth 'revolve' in elliptical orbits. This theoretical view suits astronomers because they need a predictable cyclical model for their observational work.

In fact, of course, the moon does not 'revolve' round the Earth any more than the Earth revolves 'round' the Sun. In real terms, neither travels in an elliptical orbit despite the mathematical calculations of Sir Isaac Newton and Johann Kepler which 'prove' that they do.

The moon actually travels in the form of a wave, as does the Earth. The Sun also travels in the form of a wave. The whole solar system travels in a wave form, - as do photons and electrons. This can be shown by a simple plot of the moon's path through its monthly wave 'cycle'. Thus we have Figure 5(a). The Earth's wave path is shown as a straight line although it is really curved. Figure 5(b) shows the path of the moon as it follows the Earth's wave path over twelve months. Figure 5(c) shows part of the solar system's wave patterns with the Sun in the centre.

The moon does not cycle round the Earth in equal time lapses. This can be seen to be due to various factors. The Earth's wave path is not as smooth as Figure 5(b) would suggest, it being somewhat similar to that of the moon. When the Earth's wave path changes, the moon may have travel twice on the 'inside' or 'outside' of its wave path, as it cuts through the middle of the Earth's 'S' section.

As you can clearly see from Figure 5, there is no ellipse in real terms. All the planets 'chase' the Sun, alternately overtaking it, cutting in front of it as they decelerate, slowing down until the Sun re-passes them and then accelerating again swinging behind the Sun - only to 'overshoot' and overtake it again, as they complete each 'cycle'.

Sir Isaac Newton's planetary model based on Kepler's laws would be fine if the Sun stood still. The only trouble with that is that if the Earth also stood still to allow the moon to orbit the Earth in the form of an ellipse, it is a bit difficult to see how the Earth could still orbit the Sun!! It makes me suspect that both Kepler and Newton were guilty of only using positive maths. Their model is fine for astronomers, but a bit misleading in real terms.

Language of the Universe . . . Maybe

Figure 5

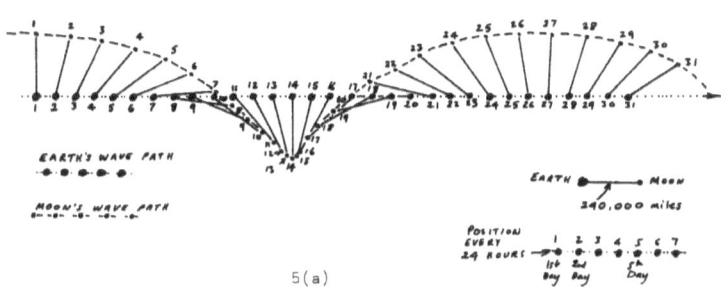

5(a)

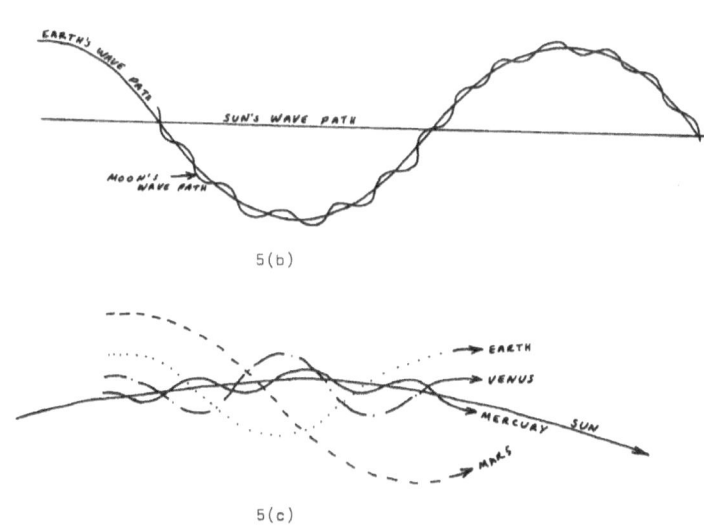

5(b)

5(c)

Chapter 14

Atomic Division

As you can see From Figure 4(a), the energy spiral would have both a positive and negative form. It would continually alternate from one to the other. This affects the atoms of which the energy spirals are a part. At any one moment of time an energy spiral can only be in one of two states: positive or negative.

Any atom depends for its charge on the relative state of the spirals within it. Thus if half the spirals were positive and the other half were negative the charge of the atom would be neutral. Positive spirals can interlock with negative ones, and depending upon the relative rotation of the spirals they will either come closer or go further away. In such cases the result could be oscillation or a constant relative bond. Spirals of the same type will repel each other.

The uni-cell organism grows to a certain size and then splits. This can now be seen from an atomic point of view based on the planetary wave patterns. As with the solar system, the smaller outer compounds are continually chasing the larger inner compounds of the nucleus. Like the moon, they pass first in front of the nucleus and then behind it.

As the cell grows, the wave paths of the outer compounds get further away from the nucleus's wave path. Eventually, a point is reached when the outer cell compounds are on a 'collision' course with the nucleus. When this happens, the nucleus splits into two, as half the outer compounds go in front of the wave paths intersection point and half go behind.

When I talk of the nucleus, I am really referring to the house and not the furniture and fittings. It is the atomic structure which divides into two new cell structures. The furniture (chromosomes etc.) then divide and join the new cells after the atomic structure has been completed. Figure 6 shows the form of atomic division.

Thus the split mechanism can be easily understood when we see the solar system in real terms from a structural point of view. Appearances can be deceptive, and this deception is one of the many paradoxes of nature.

Chapter 15

Weight and Magnetism

Another paradox concerns weight. This is another area where a true understanding of the structural realities of particle relationships is necessary if we are going to see the basic methods of cell growth.

Weight is caused by gravity. Our conception of gravity is due to the work of Sir Isaac Newton. He formulated the theory of gravity and calculated the mathematical equations which proved it. Apart from minor adjustments for relativity made by Einstein, these equations have stood the test of time and are undoubtedly valid. Despite this, Newton's conception of gravity may have been wrong.

It is my hypothesis that gravity is a force of compression rather than attraction.

My hypothesis is based on the fact that in the Universe particles have been shown to move in random directions. Our physicists call this the single symmetry principle - no particular direction of movement is favoured by the Universe. Particles are continually converging and diverging. They will continue to travel in any particular direction until they collide with other particles.

Particles in the vicinity of the Earth obey the single symmetry principle and travel in all directions. Thus any point on the Earth's surface is subjected to continual collisions caused by particles from space hitting the Earth. The interior of the Earth and the Earth's atmosphere contain more atoms per cubic foot than space. Therefore there are more collisions in the atmosphere and within the interior of the Earth.

Particles which collide with the Earth at right angles to the Earth's surface pass through less atmosphere than those colliding at an angle of less than a right angle. Thus, all other things being equal, more particles will actually collide with the Earth going straight down than if they travel in any other direction.

Some particles will pass right through the Earth, but those which pass right through the centre of the Earth are much more likely to collide with the atoms of the interior than those which miss the centre - simply because they travel further within the Earth. Thus, all other things being equal, there will be fewer particles emerging perpendicularly from the Earth than will emerge at any lesser angle.

As the Earth's atmosphere contains less atoms per cubic foot than the Earth's interior, more particles will come down through the atmosphere than will come up from the interior of the Earth.

The action of the particles colliding with the Earth creates pressure on the surface of the Earth. As the particles come from all directions, the pressure on the Earth's surface is equal in all directions. This is why the Earth is spherical.

However, there is a difference in pressure in terms of overall upward and downward movement of particles. More particles come down than go up. This pressure differential is what we call gravity.

In my opinion, objects on Earth are not 'pulled' down. They are pushed down.

Objects 'fall' towards the centre of the Earth partly because more particles hit the Earth travelling straight down and partly because less particles emerge from the Earth going straight up.

Planets which are smaller than the Earth, such as the moon, have less gravity because more particles go up from the moon's surface - because more particles pass through the moon's interior. Thus on the moon the downward pressure differential is less which means less gravity.

It seems likely that the large amount of particles emitted by the Sun in the form of light etc. is larger than the Sun's gravity. If this is so, the Sun's planets will not 'fall' into the Sun because the Sun's outward pressure would be too strong.

Two planets in space will move towards each other by compression. They are not 'attracted' to each other. Particles from space will hit the surface of each planet with equal force from all directions - except one. There will be slightly fewer particles coming from the direction of the other planet -- because the interior of the other planet will have prevented some particles from passing through. There will be slightly less pressure on the side of the planets which face the other planet. The pressure will be lowest on a line passing through the centre of both planets. Thus each planet will move towards each other's centre.

The paradox of weight concerns the strange behaviour of objects 'falling' to the surface of the Earth. For example, if we take two spheres of equal diameter - one made of lead and one made of aluminium - and we drop them from the same height at the same time, they will both hit the ground at once. However, if we then throw each sphere upwards with the same force - we find that the aluminium sphere will go higher than the lead sphere. We can throw it higher because aluminium is 'lighter' than lead.

In my view, the reason why the aluminium sphere goes higher is because it creates less resistance to the gravitational 'tide' than the lead sphere. The resistance is less because the aluminium atoms contain less sub-atomic particles such as protons, neutrons, electrons and neutrinos than the lead atoms.

The atoms or elements can be regarded as sieves. The aluminium 'atomic' sieve has a wide mesh while the lead 'atomic' sieve has a fine mesh. If you hold two ordinary wire sieves under a waterfall, the one with the finest mesh will 'appear' to be the heaviest. It is the same with atoms.

Lead appears to be heavier than aluminium because its atomic sieve creates a much greater resistance to the particles raining down on the Earth from space. The greater the resistance of an atom - the 'heavier' the atom will appear to be.

If you hold two ordinary wire sieves under a waterfall and let them go at the same height at the same time, they will both hit the bottom of the waterfall at the same moment - regardless of the difference in the size of the mesh. The downward pressure of the water completely masks the resistance differential.

In the same way, gravity masks the resistance differential of the lead and aluminium spheres. However, if the two spheres were orbited in space - at the same height and speed, the aluminium sphere would gradually outpace the lead sphere despite both of them being weightless.

The concept of 'atomic sieves' is important because it helps us to understand the mechanics of cell growth. Before turning to cell growth, I will hypothesise on my conception of the phenomena known as magnetism.

It is well known that all magnets are made of iron or an alloy of iron such as steel. The reason why iron based metals are magnetic under certain conditions is to do with their atoms.

When an iron bar is magnetised, its atoms become parallel to each other. This of itself is not significant. The important point is that the electrons of the atoms orbit in the same direction and nearly the same plane. This fact produces two effects. Firstly, particles in the gravitational stream pass easily through the longitudinal axis of the bar i.e. at right angles to the electron plane, but with difficulty through the electron plane itself. This creates a pressure differential, high pressure on the longitudinal axis and low pressure at right angles to the axis. Secondly, the particles travelling in each direction on the longitudinal axis are twisted by the electrons as they only orbit in one direction. Thus a vortex is created - clockwise at one end and anti-clockwise at the other.

When the particles leave the end of the magnetised bar, they diverge, with the outer particles arching backwards towards the low pressure area along the sides of the bar. The places where the particles converge and diverge produce the well-known lines of force.

When the particles leaving the end of the iron bar vortex into the atmosphere, they can convert iron based objects into similar magnets. The vortex has similar properties to a tornado, in that the area within the vortex is a low pressure area. Just as two planets will move toward each other because of the relatively low pressure area that exists between them, so the two magnets move together for the same reason.

If the poles of the magnets are the same, their vortexes will revolves in opposite directions. The repulsion effect is due to the interaction of the electrons within the magnetised iron bars and the other magnet's vortexes. The two vortexes cancel each other out and this forces the electrons to reverse which in turn forces the iron bar to twist and move away from the other iron bar.

Chapter 16

Cell Growth

Cell growth depends upon a number of factors, particularly the availability of food and the energy stored therein. When a cell enters a high energy environment, such as an area bathed in sunlight, its relative speed increases. The increase in speed is due to the energy field created by the photons etc., in the sunlight. The photon clouds move faster than the other elements or compounds in the area, and their presence increases the speed of anything moving in the same direction as themselves. This is because the resistance caused by particles flowing in the opposite direction is reduced.

The higher speed of the cell reduces the wavelengths of its compounds and increases their frequency. This means that the cell's sieve mesh is enlarged, and more slowly moving particles can be 'captured' by the cell.

Although the photons move everything in their path more quickly, the elements with a fine mesh sieve create more resistance and travel more slowly. The cells are largely composed of high frequency wide mesh elements like hydrogen, oxygen, nitrogen and carbon. This means they move faster and can capture the slower fine mesh elements.

Cell compounds are made up of various elements and the compounds wavelengths are the product of their constituents. Where wavelengths are similar in phasing, there is little overlap between them. By this I mean that where they overlap they are completely synchronised. Thus the mesh is larger than it would have been if they were not synchronised. See Figure 7(a) and 7(b). The greater the wave overlap, the less efficient the compound. Some elements, such as hydrogen, have wavelengths which are compatible with many others in so far as synchronisation is possible to a high degree.

The large mesh elements like hydrogen, carbon, nitrogen and oxygen remain close to the centre of the cell's wave path; while the fine mesh elements, such as sodium, calcium, phosphorous and magnesium remain further out.

The cell is very complex and therefore a fine mesh compound organism. The outer compounds of the cell are relatively of finer mesh than the inner ones. When the speed of the cell increases in the photon stream, the large mesh compounds in the energy environment outside the cell squeeze pass the outer cell compounds and get into the cell.

(A typical large mesh compound is water, composed of large mesh hydrogen and oxygen). These large mesh compounds join the cell's central large mesh compounds with the result that there is in increase in the relative velocity of the cell's centre to the cell's outside.

The increased speed of the cell increases both the distance between the outer compounds and the distance between the outer and inner compounds. Thus the cell is enlarged and relatively finer mesh compounds can enter the cell.

Figure 7

Fig. 7(a) Five 'synch' points. Five crosses. Wide Mesh.

Fig. 7(b) Two 'synch' points. Seventeen crosses. Fine Mesh

Figure 8

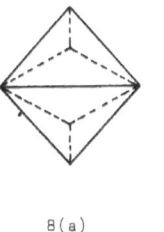 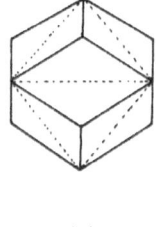

8(a)   8(b)

When the sunlight stops, the photon stream finishes and the cell slows down. The larger mesh elements and compounds then become detached from the centre of the cell and squeeze pass the outer compounds again. However, the outer compounds close ranks and trap the remaining finer mesh compounds which have entered the cell.

The increased speed of the cell increases both the distance between the outer compounds and the distance between the outer and inner compounds. Thus the cell is enlarged and relatively finer mesh compounds can enter the cell.

In this way the cell gradually 'grows' until the outer compounds are so far from the centre compounds that they reach a 'collision' wave path. The cell then splits in the manner described in Chapter 14.

In my view, elements can only join up with other elements to form compounds if there is a degree of synchronised overlap somewhere on their collective wave frequencies. A complete lack of any synchronisation will mean incompatibility. Frequent synchronisation will mean good bonding. Elements which are by themselves incompatible may become compatible by joining compounds containing the incompatible element. The joint wave length may allow synchronisation to take place.

So the acceptance of elements or compounds into the cell structure depends on their wavelength compatibility. Water can become a carrier compound for many elements because of the wide degree of synchronisation offered by its compound wavelength. Such elements are said to be 'soluble'.

Chapter 17

Cell Joins

It is my hypothesis that uni-celled organisms are slowly reacting groups of chemical compounds which reproduce themselves by splitting, by a double reversal process. These uni-cells have a definite identity due to their chemical compounds and this gives them a specific compound wavelength pattern.

It is because of this multiple wavelength, due to the large number of compounds which make up the uni-cell, that such cells are normally incompatible with other uni-cells. This is partly because their identity varies according to their age. The age of a cell indicates the position in time that it has reached in its overall chemical reaction.

When a cell splits, its wavelengths are incompatible because they are separated both in space and time. In addition the uni-cell has a certain overall charge due to the positive and negative energy spirals within the cell compounds. The charges of the newly divided uni-cells are therefore identical and the new cells repel each other. Of course, given millions of such uni-cell organisms and millions of years, it is possible that two uni-cells which were absolutely compatible in both their wavelength identity and their overall charge would meet. It is my hypothesis, regarding cell joins, that this happened in the past, and that the result was a 'double-celled' organism. In such a double-cell, the two nuclei would not merge into one, because this would immediately put them on a collision wave path due to their combined size. Their compatibility would prevent them from so merging and splitting. They would be compatible, not identical.

The two compatible uni-cells would remain in the form of a double-cell as long as each remained compatible both in terms of charge and wavelength pattern. If my structural hypothesis on the tetrahedron shape of uni-celled organisms is basically correct, as per Figure 3(d) in Chapter 11, a double-celled organism would form a trigonal bi-pyramid shape as in Figure 8(a). This reverses into a rhombic duo-decahedron as shown in Figure 8(b).

It is interesting to note that the cells of most multi-celled organisms, whether plants or animals, conform to the rhombic duo-decahedron shape.

The double cells would grow in exactly the same way as the uni-cells, - indeed they would probably grow better because of their wider range of wavelengths. Growth would be very even due to the inter-linked nature of the double-cell, and the two uni-cells making up the double-cell would split at virtually the same time.

After splitting, the 'parent' uni-cells would remain inter linked, because they would still be compatible. The other half of the split uni-cells would not be compatible with their own 'parent', but they would be compatible with their opposite half and the opposite 'parent'. The 'new' uni-cells would therefore inter link with each other, after crossing over so that they remained next to their opposite 'parent'. Thus the double cell would become a quadruple cell, after binary fission, consisting of two pairs of inter linked uni-cells sticking together side by side. Two double-cells side by side. A small number of splits would produce a large multi-celled organism.

Chapter 18

Double-cell Implications

The concept of double cells fits quite well with known facts on cellular constitution. Most cells in multi-cellular organisms are of this type, known as 'diploid' cells to the biologists. Human cells are an example of the diploid cell, their nucleus's containing double chromosomes which are intertwined around each other.

The double cell configuration projects certain structural and electromagnetic implications. Unlike uni-cells, the double cell is not Omni-directional, and any multi-cellular growth is affected by this fact. Furthermore, if the double cell is to grow and reproduce by binary fission, it must have access to a food source and reproduction space. These requirements restrict the sort of growth patterns which can be achieved.

The double cell based organisms can only grow in the following structural forms:

Solid tubes, which consist of double-cells joined end to end, as per Figure 9(a).

Sheets, consisting of double-cells stacked together either vertically or horizontally, as per Figures 9(b), and 9(c).

Hollow tubes, consisting of sheets of double-cells which have joined edge to edge, as per Figures 9(d), 9(e) & 9(f).

Apart from the structural implications due to tine double celled configuration, there would also be electromagnetic implications. Each double-cell would consist of a positively charged uni-cell inter linked with a negatively charged uni-cell. Figure 9 shows the positive and negative parts of the double-cells divided by dotted lines. The drawings reproduce the double-cells as though they were double cubes. However, as I have said in Chapter 17, the double-cell base is likely to take the form of a rhombic duo-decahedron. This means that the cells do not line up in a rectangular form, but adopt a diagonal/hexagonal layout as per Figure 9(g). As a result, the cell structures have sides or edges which are all positively or all negatively charged.

The effect of the electro-magnetic configuration of the overall multi-cellular organism, whether solid tubes, sheets or hollow tube forms, is determined largely by the environment of the organism. The main determining factor is the Earth's surface which carries a negative charge. As a result the solid and hollow tubes tend to grow in a vertical plane. Sheets vary according to their configuration. Some will grow in the vertical plane while others will prefer the horizontal plane.

Algae may be examples of solid tubes, while leaves may be examples of sheets. Hollow tubes are structurally very strong and most plants and animals of any size are likely to be constructed out of such tubes. As you can see in Figure 9(d), (e) & (f), the hollow tubes can come in a variety of structural configurations. There could be tubes within tubes as long as the double-cells had one of their sides exposed to a food source. The double-cells would only split by binary fission if there was room for such reproduction.

Language of the Universe . . . Maybe

Figure 9

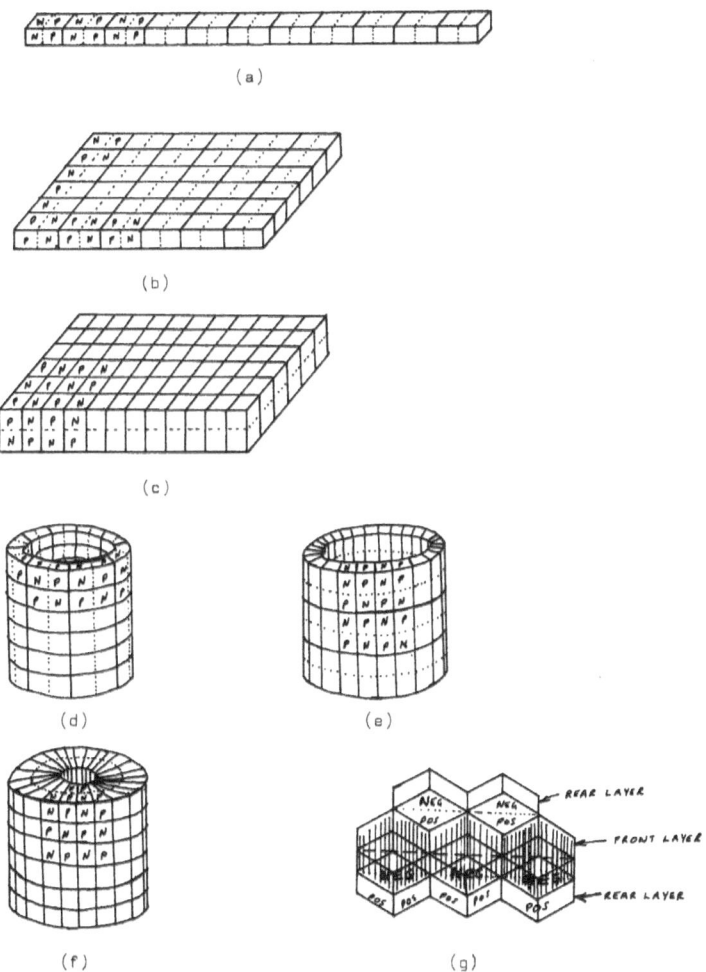

Chapter 19

Sexual Reproduction

In my view, sexual reproduction is a result of the double-cell configuration and developed with the multi-cellular organisms. Thus the multi-cellular organisms have two forms of reproduction whereas the uni-celled organisms only have one. The multi-cellular organisms can split by binary fission like the uni-celled organisms.

They can also spilt by a separation of their double-cells into two uni-cells and these uni-cells can recombine with other compatible uni-cells.

The biologists call the unicellular binary fission split - mitosis, and the double-cell separation - meiosis. The recombination of the separated uni-cells is called fertilisation.

Of course, the biologists do not see the double-cell as such, but rather see it as a single cell. In their view this single cell simply reduces its chromosome content by half during meiosis. It is then regarded as a half cell, or 'haploid' cell.

It is my contention, as you have observed, that the uni-cell structure is hollow and allows for the replication of its structure by this fact. The double-cell therefore consists of two 'hollow' uni-cells intertwined. Just before mitosis, the two cells have replicated themselves becoming four cells - two internal and two external. Slightly different from the current biological view

Language of the Universe . . . Maybe

In this chapter I will consider sexual reproduction as conducted by a multi-cellular organism which has grown in the form of a hollow tube. The same basic system of reproduction would apply to all the other structural forms previously illustrated.

The hollow tube stands vertically on the Earth's surface with its positively charged end attached to the negatively charged Earth. The hollow tubes upper end is negative. It is repelled by the Earth, and occupies a position as far away from the Earth as possible -that is -- vertically above the base of the tube.

The organism consists of sheets of double-cells wrapped up in a tube form. Each layer of double-cells being separated by a food transport medium such as a layer of water.

As the organism grows by mitosis, starting from an original double-cell, it follows that those original cells must be located at the structural centre of the organism. When the original double-cells separate, because their wavelengths or electromagnetic charges are no longer compatible, the positively charged 'half' would be attracted towards the negatively charged Earth and would go down; while the negatively charged 'half' would go up the tube.

The separation of the double-cell would take place only when the wavelengths became sufficiently out of synchronization to overcome the bonding caused by the wavelength compatibility, or the electromagnetic compatibility.

Alternatively, the electromagnetic charges might become sufficiently identical to cause the separation, but they would have to overcome the bonding effect of the wavelength compatibility first.

This double retention aspect of double-cell inter linking would mean that the double-cells would separate long after the mitosis splits. Once they had started to separate, however, the separation process would continue, cell by cell, until the whole organism disintegrated. Only continuous mitosis would prevent the separation process from breaking down the multi-cellular organism. If mitosis stopped, due to the end of the chemical reaction process of the uni-cells, or because of lack of food, the meiosis separation would eventually overtake the organism, and it would disintegrate.

So, to continue with the structural implications of double-cell separation within a hollow tube, I draw your attention to Figure 10(a), which shows a typical tube. The original double-cells are at the centre, halfway up the tube and on the inner surface. The negative half goes up and the positive half goes down.

The positive halves of the separating double-cells would clearly be trapped in the bottom of the tube. The negative halves would also be trapped temporarily at the top of the tube - because the negative top end of the tube would repel them. Eventually the negative halves would overcome the repulsion effect and would float out of the top of the tube.

Given large numbers of such tubes, any negatively charged halves, which were compatible with the positively charged halves trapped in the tube bottom, would be attracted to the overall positive charge emanating from the trapped uni-cells.

Initially, the compatible negatively charged halves from the other tubes would be unable to overcome the repulsion effect of the negative 'valve' consisting of the top of the tube. However, when a large enough number of negatively charged compatible uni-cells had gathered at the top, one of them would break though. This one would go down into the tube. It would attract the top most of the positively charged halves, which by virtue of their large numbers and similar charge, could not all be at the bottom of the tube.

The top-most one would therefore meet the negatively charged uni-cell about three quarters of the way up the tube. Thus we have Figure 10(b) which illustrates this meeting.

The two compatible uni-cells would then form a new double-cell. In some organisms the double-cell might float out of the top of the tube, but most would probably develop inside the tube, near the top, until it got too large - whence it would be repelled by the 'parent' tube.

This is the basic form of sexual reproduction as I see it.

Language of the Universe . . . Maybe

Figure 10

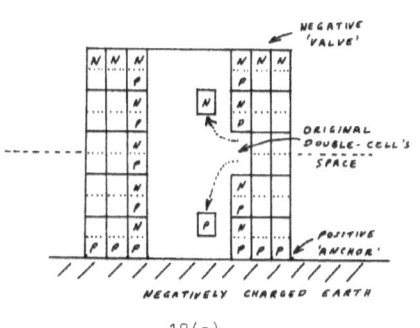
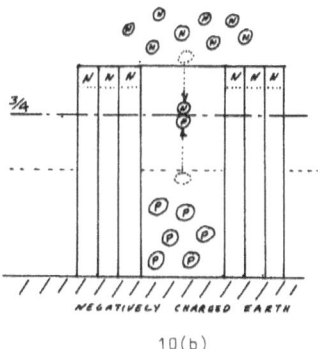

10(a)   10(b)

Figure 11

'A' Type   'V' Type   'H' Type

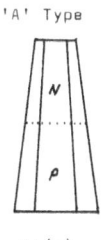   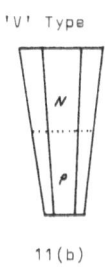   

11(a)   11(b)   11(c)

Chapter 20

Sexual Effects

The structure of the double-cell is determined by its chemical composition and this dictates whether it takes the form of a solid tube, sheet or hollow tube. Each double-cell will replicate its own structure until one of the multi-cellular forms is produced. Changes in chemical constitution would lead in turn to changes in the shape of the multi-cell structure.

The structural evolution of multi-cell organisms, suggests that the first form was a solid tube, then came a sheet and finally a hollow tube. Each structure pattern may represent a different stage in the overall chemical reaction time of the organism. Thus, the more complex the chemical constituents of the organism, the longer the overall reaction time would be, and the greater the likelihood of pattern changes, due to overall changes in chemical constitution, at any particular time.

This is particularly noticeable in large multi-cellular organisms such as trees. The first form taken is that of a solid tube. This may develop into a sheet i.e. a leaf. Alternatively it may develop into a hollow tube, i.e. a sheet joined edge to edge. I suspect an organism's cells will grow in the form dictated by the timing of their chemical reactions. The 'life' of a cell is the time taken for it to complete its overall chemical reaction. When the reaction is complete, the cell becomes inanimate. During its overall reaction it may grow and split by binary fission many times.

I would surmise that leaves are formed by cells which have reached the stage of 'life' where their chemical composition has altered from that which they had in their solid tube form. Those cells which can expand into sheets by virtue of being at the end of the tree (hollow tube), and therefore having growth space, will do so. The cells underneath them, although similar in age, cannot expand and remain as part of the hollow tube structure of the tree. Later on, after the leaves have been produced, although the cells which were underneath now have growth space, they have changed their chemical composition because they are older, and they grow in the form of a hollow tube.

As I have said, it is my view that all advanced multi-cellular organisms developed in the hollow tube form. This includes trees and animals. Thus trees have a hole in the top and a hole in the bottom, or rather a very large number of holes in the top and bottom. Holes in their leaves and holes in their roots. Animals of course, have holes in their heads and holes in their bottoms. They are all forms of hollow tubes.

By virtue of their ancestry in a vertical hollow tube form, all animals (including insects) have separating double-cells - some of which leave their bodies and others which stay trapped inside. It is my contention that these trapped uni-cells are responsible for the changes which occur during adolescence. In other words, the hormones are triggered by these separating double-cells which cannot get out, and which build up either a positive or negative charge in the animal's body.

Chapter 21

Heterosexual Reproduction

One of the main effects of the vertical hollow tubes relates to stability. Both the influence of the Earth's negative charge and the environment of the multi-cellular organism would tend to favour tubes which were both electromagnetically and structurally stable. Hollow tubes anchored to the Earth would not survive unless their environment brought food to them. This means the environment would have to be mobile, consisting of either a liquid or a gas.

Such conditions would favour tubes which had a large base and small top, and which were positively charged at the bottom of the tube.

The growth pattern of such tubes would in any event depend on the double-cell alignment. Thus cells arranged as per Figure 9(e), would grow much faster upwards than they would outwards. The result would be a long tube with a thick base and small apex. The base would be strongly positive and the apex weakly negative.

The horizontally aligned double-cells of Figure 9(f), would produce something which was fat in the middle, with a lot of spherical growth. This would tend to look like an onion or a bulb.

The other horizontally aligned double-cells of Figure 9(d), would produce a much more even cylindrical structure with a slightly smaller apex than base.

The electromagnetic effects would have to be much more subtle. The reason for this is that if a multi-celled organism had double-cells with a very large positive or negative bias, the resultant growth would be completely spherical rather than flat and sheet like. A spherical form would not allow the inner cells to feed, and the organism would die.

Thus the differences in the double-cell configuration would have to be very slight. The combination of double-cells would always tend towards the average. The double-cell structures can he grouped into three types as per Figure 11(a), 11(b) & 11(c).

The 'A' type combination would be the most stable. The 'H' types would be the next most stable, while the 'V' types would be comparatively unstable. However, the 'H' types, although average, would be rare. Most hollow tube based organisms would be 'A' types or 'V' types. This is because in any sample there are usually only a very small number of absolutely average types, with the greater part of the sample tending towards one extreme or the other.

The 'V' type's lack of structural stability would tend to encourage it to turn upside down. Although it would not be electromagnetically stable - its base being the same charge as the Earth - it would be more structurally stable and this would be more important in a mobile liquid or gaseous environment.

Thus we get the sexual reproduction pattern shown in Figure 12(a). The positively charged uni-cells of the 'V' type structure are attracted to the Earth. Thus they tend to leave the tube more slowly than the type 'A' negatives. As the dotted line shows, the type 'A' negative would tend to combine with the type 'V' positive, and then the new double-celled organism would leave the type 'V' parent. This, in my view, is the basis or heterosexual reproduction in plants and animals.

Animals are hollow tubes in 'reversed' form. By reversed, I mean that while plants mainly get their food via their bottom holes (roots), and excrete and reproduce via their top holes (leaves and blossoms); animals eat through their tops, and excrete and reproduce through their bottoms.

Thus we get the reversed animal arrangement, as per Figure 12(b). It is my hypothesis that the hollow tube ancestry of animals has left them with slight differences in their structural configuration which leads not only to a different shape, but also to different sex roles.

Figure 12(b) suggests that type 'V's are normally male, while type 'A's are normally female. It is clear, however, that the sex of an individual organism is not dependent on its structure, although its structure may affect its sexual success.

Language of the Universe . . . Maybe

Figure 12

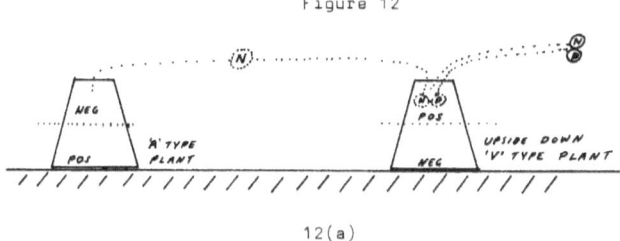

12(a)

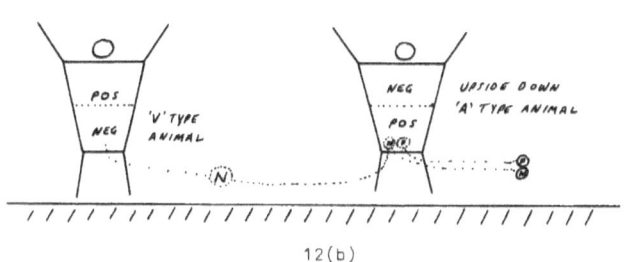

12(b)

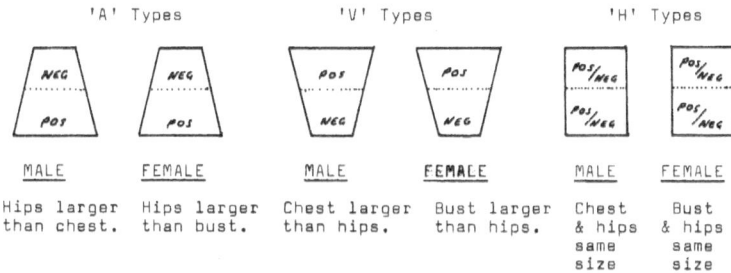

Figure 13

Chapter 22

Human Archetypes

As humans are animals, they inherit the electromagnetic and structure typing of animals. Thus humans have two basic 'archetype' and an average 'archetype', from a structural point or view. Although the species continually works towards the average, the average archetype is comparatively rare. There are male and female forms of the archetypes, as illustrated in Figure 13.

Nearly everybody in the human race is either an 'A' type or a 'V' type. Few people are 'H' types. The compatibility or the types and their electromagnetic charges affect their likely combinations. Thus the 'V' type tends to find the 'A' type attractive and vice versa. Both the 'V' and 'A' types will find the 'H' type attractive and the 'H' type will find them equally attractive.

'A' types will tend to repel other 'A' types and 'V' types will similarly repel other 'V' types. 'H' types will tend to equivocate towards other 'H' types, being alternately attracted and repelled.

The 'H' type male and female should be the most sought after type as they attract all the other types. It is interesting to note that the "Miss World" contestants are usually female 'H' types with their vital statistics almost invariably close to 36" - 24" - 36". By contrast, the "Mr Universe" contests are for male 'V' types only, and as such the contest is not so popular. 'H' type pop stars and film stars can usually maintain a much larger female following.

It would be interesting to know whether the configuration of a 'type' affects the way people see things. Would 'A' types looking at a square, tend to see the bottom of the square as longer than the top? If they did, they would presumably also tend to see 'V' types as average. Only 'H' types would see people as they really were, if this was the case.

The effects of the electromagnetic and structural configurations on personality would also be interesting. I would predict that 'A' types would be stable but dogmatic. 'V' types would be unstable and explosive. 'H' types would be contrary, equivocal and indecisive.

I have tabulated my predictions for the 'archetypes' in appendix B.

Chapter 23

Sexual Selection

At this point, I will consider the current biological view that the sex of a baby is determined on a chance basis. This view is really rather strange. As with planetary motion, it looks as If the biologists have not considered the implications and contradictions of their views. Of course, religious considerations originally enforced the current view, and philosophical reasons may have prevented a close examination of the problem.

That a child's sex should be determined by chance flies in the face of common sense, mathematical probability and logic. It is my contention that the sex of a child is determined largely by the electromagnetic and structural configurations of its parents.

Let us consider a few facts. Most advanced multi-celled organisms such as trees, insects, fish, reptiles, birds and mammals, do not engage in mating until they have reached their adult form. During the organisms' growth to adulthood, they protect themselves with many defence mechanisms, to ensure that they stay alive long enough to mate. Thus they take no chances with their survival up to mating. I repeat, nothing is left to chance.

The survival of a species depends not only upon adulthood being reached and mating taking place. It also depends on the reproduction of new organisms which are evenly balanced, -- structurally, electromagnetically and sexually. Failure to maintain an even sexual balance could mean the extinction of the species.

Consider, would an organism take an enormous number of steps to defend its survival to adulthood and then leave the most important element in its specie's survival purely to chance? Mathematical logic suggests that if left to chance it would be quite possible for one sex only to be selected every few hundred generations. If this happened the species would die out. It doesn't happen.

Consider again, nearly all the two million or so known species of heterosexual organisms produce an extremely finely balanced brood, generation after generation, and have done so for millions of years. Think of it in terms of 'Roulette'. It would be like betting on 'evens' on two million roulette wheels and having the ball fall on an even number for millions of turns on the run without a break. Now is that likely? It is possible, but very, very, improbable. Nevertheless, this is the current biological view.

Other characters are determined according to certain genetic rules, but apparently the most important character of all is left to chance.

Well if it is not chance, then how does it work? It is generally agreed that the X and Y sex chromosomes in humans are responsible for the sex selection of a child.

It is my hypothesis that these sex chromosomes carry electromagnetic charges. The

'X' chromosome would have two negative and two positive charged arms, while the 'Y' chromosomes would have either one positive and two negative arms, or one negative and two positive arms. This variation in the charge of the 'Y' chromosome would be responsible for the sex selection or the child.

In my view the two uni-cells combine structurally so that the 'X' and 'Y' sex chromosomes are directly in alignment. I envisage a female uni-cell in reversed form i.e. half a rhombic duo-decahedron, joining a male uni-cell which has an unreversed tetrahedron shape. Thus the male uni-cell actually enters the female uni-cell as in Figure 14(a).

The inside end of the female uni-cell would take the rhombic duo-decahedron's molecular X form, while the tetrahedron tip of the male uni-cell would take a Y form. After entering the female uni-cell, the male uni-cell would orientate itself according to its charge. It would then reverse itself into the other 'half' of the rhombic duo-decahedron, forming a complete double-cell. Thus we have Figure 14(b).

The relative alignment of the male uni-cell would dictate the electromagnetic structure of the new double-cell and this would account for the difference in male or female growth patterns. Thus the male parent's 'Y' sex chromosome would decide the sex of the child and the chromosome would vary with the charges on its arms.

Language of the Universe . . . Maybe

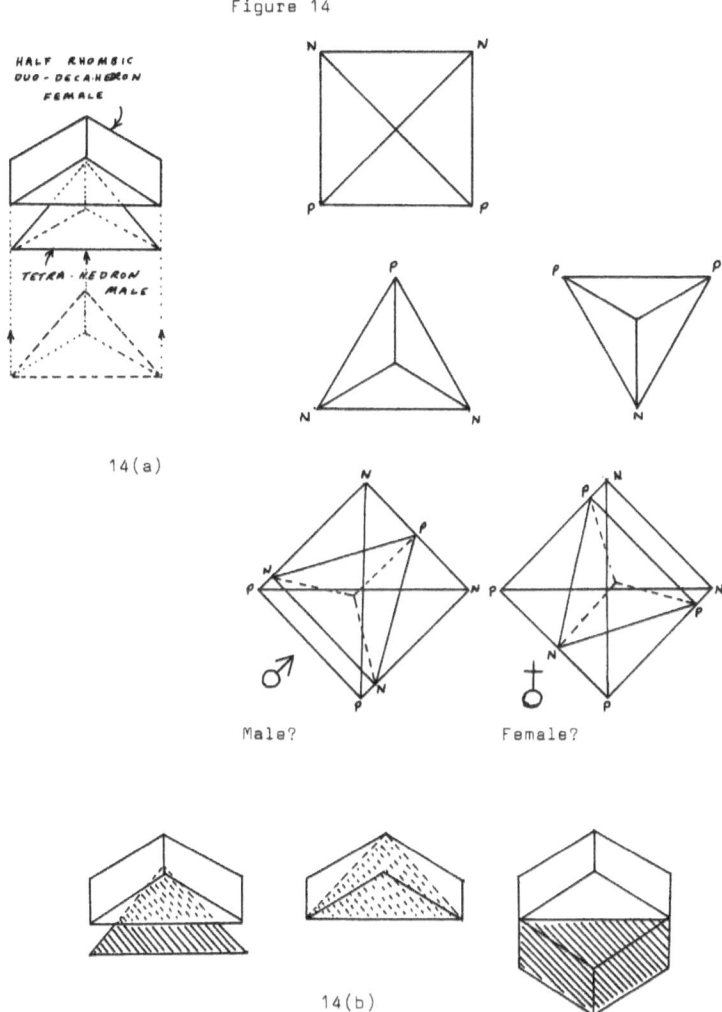

Figure 14

Chapter 24

Progeny Guide

In my opinion the structural typing will give us the main clue as to the likely sex of the progeny of an organism. If we go back and look at the apes, we can see the 'types' and their progeny.

In Miss Goodhall's Gorme Stream study, the high ranking males were invariably athletic and strong - almost certainly 'V' types. The timid females would probably be 'A' types. The timid females give birth to more sons than daughters. The sexually aggressive females mate with the low ranking males and produce more daughters than sons.

The low ranking male is probably either an 'A' type or a 'H' type, while the sexually aggressive female is probably a 'V' type or a 'H' type. (Miss Goodhall didn't supply any vital statistics). We can think of female film stars and actresses, who are aggressively sexual and invariably seem to have bigger busts than hips. As such they are 'V' types. If they marry a 'V' type film star they rarely seem to have any children. They tend to be more successful, family-wise, if they marry a pear shaped 'A' type.

The average 'H' types will tend to have more daughters if married to an 'A' type, or more sons if married to a 'V' type. If the species is to maintain a balance, the progeny must tend towards the average, both sexually and structurally. Figure 15 indicates a possible guide to the progeny of various types.

The last two matings in Figure 15 seem fairly improbable and would probably give rise to mating difficulties. 'V' type females, if slightly built, would be expected to have a relatively narrow pelvis, which would make mating and child bearing more difficult. The 'V' type males could be expected to be more successful than the 'A' type males because of their compatibility with the good mating and child bearing female 'A' types and 'H' types.

These structural implications make a feedback loop to the interpersonal behaviour patterns. The relationship between parents and their children would appear to vary according to their structural compatibility. Thus 'A' type mothers would electromagnetically repel their 'A' type daughters, if they had any -- particularly after adolescence when the negative uni-cells build up. Similarly, 'V' type fathers and sons would be at loggerheads. 'V' type fathers would get on well with any 'A' type daughters; and 'V' type sons would see eye to eye with 'A' type mothers.

The 'H' type children would create problems for 'V' and 'A' type parents. The children would be like neither parent, although both parents might find such children attractive.

In a large family, there might typically be an 'A' type mother and several sons. If one of the sons was an 'A' type, while the other sons were 'H' or 'V' types, the mother would probably continually find fault with the 'A' type son. In such a family, the 'V' type sons would probably be the mother's favourites. In this way, children can be temperamentally very different from each other; and the parents will select favourites without really knowing why.

Thus the inter-relationships of the structural influences and the behavioural responses begins to take shape. It can be seen at once how a very few, very simple and basic differences can quickly multiply into an enormous number of behavioural variations.

Figure 15 - Prodgeny Guide

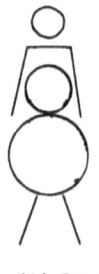
'A' Type

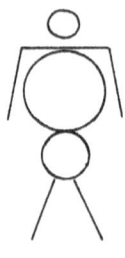
'V' Type

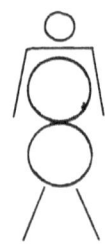
'H' Type

MALE TYPES	MATE	FEMALE TYPES		RESULTS
'A' type	X	'V' type	=	mainly 'H'&'V' daughters
'V' type	X	'A' type	=	mainly 'H'&'V' sons
'A' type	X	'H' type	=	mainly 'A' daughters
'V' type	X	'H' type	=	mainly 'V' sons
'H' type	X	'A' type	=	mainly 'A' daughters
'H' type	X	'V' type	=	mainly 'V' sons
'H' type	X	'H' type	=	even 'A' sons and 'V' daughters
'A' type	X	'A' type	=	Problems?
'V' type	X	'V' type	=	Problems?

Chapter 25

Environmental Evolution

My next hypothesis concerns the evolution of organisms due to their environment. The Darwinian view is that organisms change according to the fitness with which they can survive in any given environment. Thus those species which can adapt to environmental changes will outlive those which cannot so adapt.

My view is that organisms also change at the same time as the environment and that their evolution is part of the environmental evolution. So organisms have developed chemically and structurally as the rest of the environment has evolved. Organisms are chemically evolving reactions just like other aspects of the evolving universe.

The inter linking of organisms to their environment means that the structural shape they assume is that which best suits the external environment. When thinking of most structural growths, (like crystals for example), we tend to think of them forming without any influence from their external conditions. We say that crystals grow to the shapes they form because this is the way their atoms are arranged.

What we don't so often say is - that their atoms are so arranged because the neighbouring environment forces them into that shape. We tend to assume that they form -- as it were -- in a vacuum. Darwin appeared to assume that organisms were somehow separate from the rest of the environment. In my view, they are an integral part of it -- part of the evolution of the universe.

Turning from general concepts to specific ones, I would like to look at multi-cellular organisms. As I have said previously, most animals and plants are based on a hollow tube structure. Trees may be said to have a vertical axis, while grazing animals and fish etc. have a horizontal axis. It is my hypothesis that differences between classes of species are due to structural causes, while differences within a class of species are due to minor differences in the chemical compounds of their cells.

In terms of structural shapes, I think that the double-cell has within it a number of divisions. The arrangement of these divisions is responsible for the axial orientation adopted by the organism. The divisional arrangements will have an optimum orientation. The species which most closely meets that optimum orientation will be the most successful from an environmental compatibility (and therefore survival) point of view.

I think it is likely that mutations which improve a cell's orientation to the optimum will succeed, while contrary mutations will fail. Thus the evolving organism will gradually develop until its structural shape is synchronised with the optimum orientation of the cell's divisional arrangements. However, after this point, the evolution of the cell must be divisional rather than orientational. Each divisional change will mean a different optimum orientation; and the species will slowly change -- by mutation -- until it has conformed to the new optimum orientation. Then there will be another divisional change.

If my hypothesis is correct, then it means that older classes of species are just as important as the newer ones. This is because the older classes of specie provide part of the evolutionary background for the new ones. The old classes of specie are responsible for producing the environment conditions which the newer ones need to evolve within. Thus the extinction of the older classes of specie may mean the eventual extinction of the newer classes of specie which have evolved from them.

Old classes of specie may be responsible either for producing new chemical compounds or elements; or absorbing elements or compounds -- which if left unabsorbed would destroy the newer classes of specie.

Chapter 26

Evolutionary Time wave

The evolution of species due to changes in their divisional arrangements can be charted - and it is interesting to see that they appear, in terms of optimum orientation, to conform to a wave pattern. In other words, time can be seen to move in a wave form within evolution as within other forms of energy motion. Evolution is very slow but it still appears to conform to the same structural pattern as faster forms of motion

Figure 16 shows the 'evolutionary time wave' in action, as each species gradually conforms to the optimum orientation of its cellular divisional arrangements. It will be noted that all species are on the main line of evolution. Here, of course, I disagree with the current view of most natural historians who put birds on a side branch of the evolutionary 'tree'. The evolutionary time wave and the 'waves of motion' in the next chapter, put birds clearly in the main stream.

It is interesting to see penguins at the end of the evolutionary time wave for birds.

This means that I think penguins are the most successful birds from an environmental survival point of view and the last form of birds before the marsupials.

When I originally worked this out, I couldn't believe I had got the right answer. After all, how could penguins possibly be the world's most successful bird when it can't even fly? The penguin's lack of flying ability is surely one of nature's supreme paradoxes. However, when you consider the matter of environmental survival, the penguin's place at the top of the tree becomes obvious.

The penguin can live on land or sea. It can live on the equator and can winter out on the South Pole. It is thus supremely adaptable. If the world got a lot hotter than it is now, the penguin would survive as well as any bird which lived in the equatorial regions, and better than land birds, because of its ability to live in the cooler sea water.

If the world got extremely cold, and there were six months of Antarctic temperatures to endure, only the penguin would survive. If birds needed to fly all the time, they would all die because they could not mate or rear their young in the air. Thus the penguin's lack of flying ability is no real handicap from a survival point of view.

I think the penguin developed into something like an echidna. The echidna is an Australian ant-eater which is considered to be half way between birds and marsupials. The echidna would then develop into a marsupial such as a wombat. It is noticeable that penguins have brood pouches where they incubate their eggs for a very long time. The marsupials also have pouches, and their young, although born alive, are very small and gradually grow up in the pouch.

The current view held by many natural historians holds that the mainstream of evolution went straight from a four legged reptile (now extinct) to a dog-like mammal. This does not conform either to the evolutionary time wave or to the development of waves of motion.

If my hypothesis is right, each class of specie can be expected to have its moment of glory when it will be the most abundant class and the most successful from an environmental survival point of view. At this time it will develop into its most varied forms. Later, as it declines, only the most successful examples of the class will survive, and they are likely to be the species which most closely represent the class's structural position on the evolutionary time wave.

Language of the Universe . . . Maybe

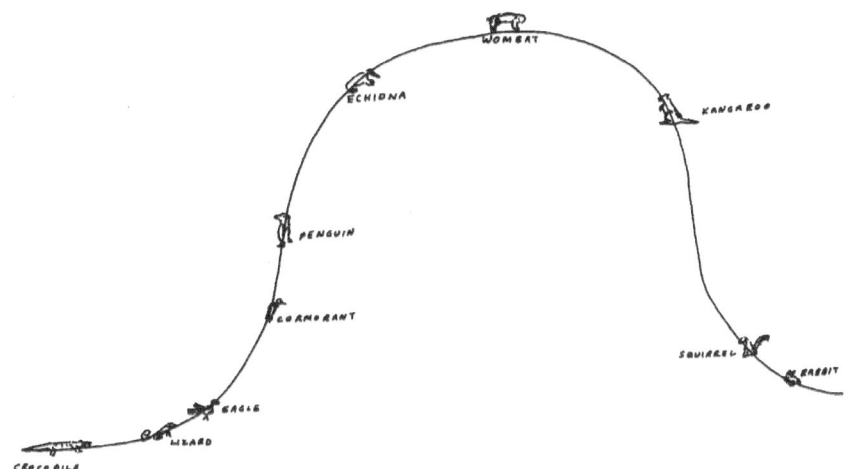

Figure 16 Evolutionary Time Wave

Chapter 27

Waves of orientation and motion

In the first part of this chapter, I will discuss my hypothesis that the divisional arrangements inside the double-cell cause the shape and orientation of the relevant species. The hypothesis assumes that all the divisions are equal, that the compartments are either positively or negatively charged, and that they are so arranged that they cannot be divided without the total disintegration of the double-cell. Because of the double-cell, each division comprises a pair of compartments joined together. The compartments build up in the form of prime numbers

We start off with three compartments. This would be like three golf balls joined to each other. There is an axis through the centre and the compartments are charged as follows:- one positive, one negative, and one alternating from positive to negative. The alternating charge is not very efficient, and it needs a lot of energy to maintain the charge. Thus we have Figure 17(a). It is interesting to see the similarity of this structure with that of insects, and to note that insects live almost entirely on a very high energy food, namely sugar.

Next we have five compartments as per Figure 17(b). The axis cannot go straight down the middle. It must go in from one side and come out on the same side. The charges would be three positives and two negatives. This produces a structure like that of a shark.

Next we have seven compartments as per Figure 17(c). The axis again cannot go straight through. It enters under the front compartment and exits under the 'tail'. The charges are as per the drawing, and the structure is like that of a typical bony fish

Then we jump to eleven compartments as per Figure 17(d). This produces a reptile like structure. Thirteen compartments are next, as per Figure 18(a). Behold - a bird

Next we jump up four more, to seventeen compartments as in Figure 18(b). My kingdom for a horse! At nineteen we see a lot of monkey business and at twenty three….

Work it out for yourself.

The hypothesis may be a load of nonsense, but you've got to admit it looks good.

On Figure 19, we move on to the waves of motion. This drawing illustrates the evolution of movement of the species. The species continually oscillate from single wave propulsion to twin wave propulsion.

J.D. Waters

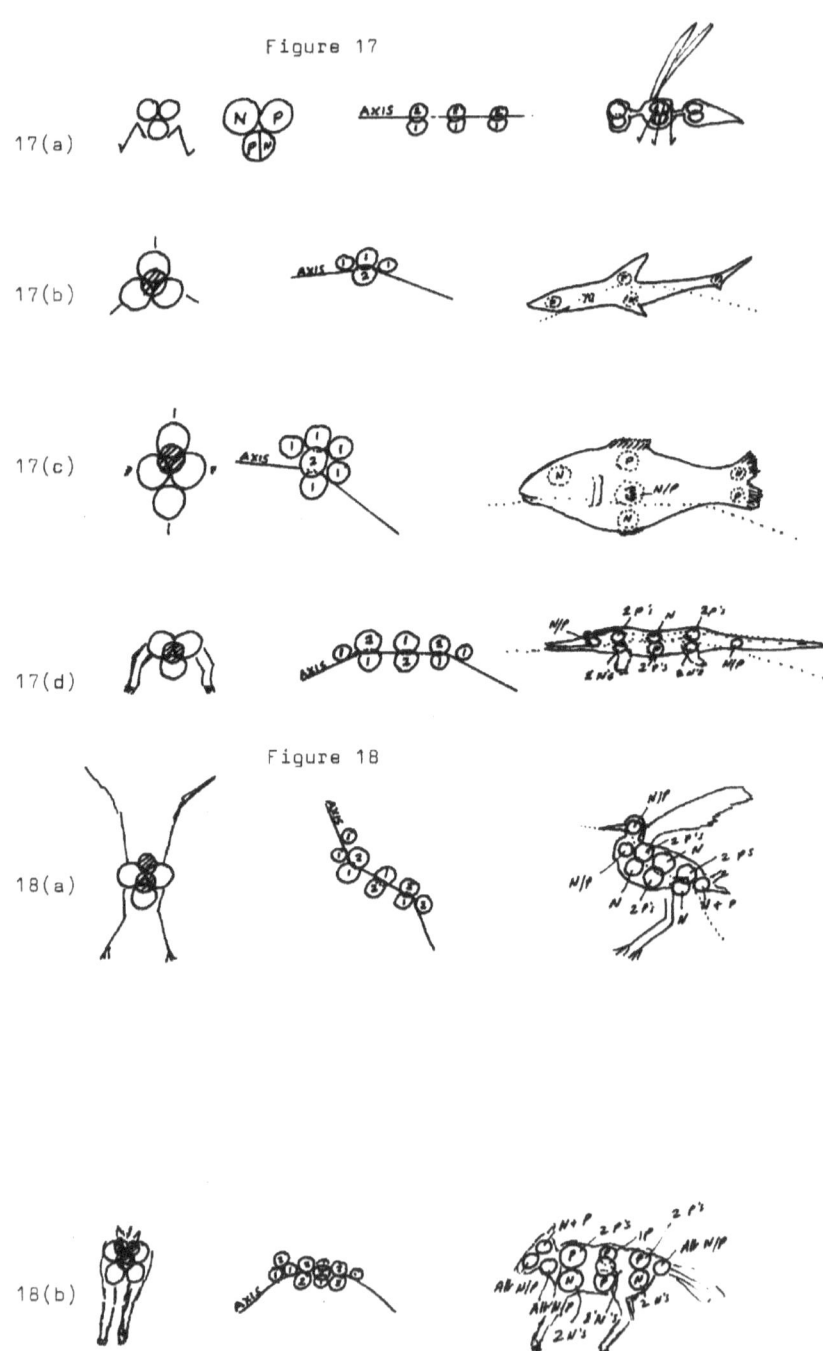

Figure 17

17(a)

17(b)

17(c)

17(d)

Figure 18

18(a)

18(b)

Language of the Universe ... Maybe

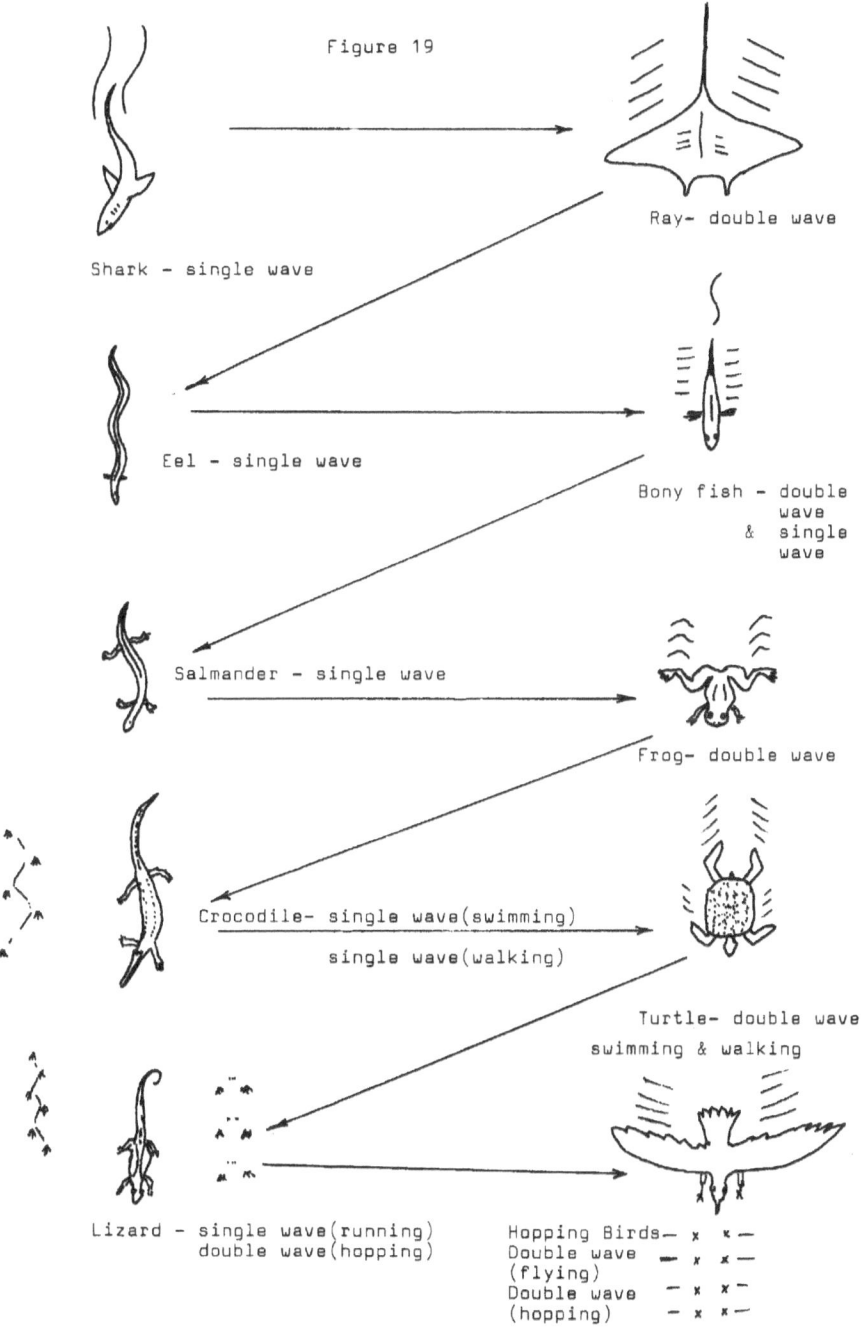

Chapter 28

Mendel's Peas

In his book "The Ascent of Man", Jacob Bronowski wrote about Father Gregor Mendel. Mendel is considered to be the father of genetics, and worked out the laws of genetic inheritance after years of experiments with the sweet pea. When he carried out his experiments he decided to test for seven characters. It just so happens that the sweet pea has seven chromosomes; and if Mendel had selected six or eight characters to test for, his experiment would not have succeeded. Mendel picked the right number first time.

Now Bronowski wondered whether Mendel's choice of seven characters was just a matter of luck, or whether Mendel had determined the right number by some unknown means, before he started his experiments. The very simplest answer to this question may not be the correct one, but surely it is worth considering.

Gregor Mendel was a monk who lived in an Abbey in Brno, Czechoslovakia. Monks grow their own food, prepare their own meals and pray. Mendel must have spent many happy hours peeling potatoes, dicing carrots, shelling peas and praying. When shelling the peas, he must have noticed that each pod contains the same number of peas.

Now that's not the sort of thing Jacob Bronowski would probably have noticed, I expect he only ate frozen peas anyway. But if you had to conduct a lengthy series of experiments, to determine the number of characters passed from a parent pea to its seeds (the peas in the pod); and you had to pick a number; and you knew there were always seven peas in a pod, which number would you have chosen?

This coincidence or number of peas and number of characters may be misleading. Does the number of seeds in a pod have any significance? Does the structural arrangement of the seeds matter? In most seed producing organisms, both the number and the arrangement or the seeds is repeated again and again. I wondered whether the arrangement of the seeds had anything to do with the arrangement of the chromosomes within the cell, or with the internal structure off the cell itself.

Chapter 29

Crystal Trees

While I was wondering about the relationship of the structure of the cells to the arrangement of the seeds, I decided to look more closely at the external structure of organisms to see if they reflected any simple molecular shapes.

An examination of trees showed that they invariably seemed to follow a hexagonal shape. The conifers were less regular than the broad-leaved trees in outline. The external outlines could be seen most easily in the summer when all the leaves were out. Thus we have the tree outlines illustrated in Figures 20(a), 20(b), 20(c), and 20 (d).

I wondered why all the ends of the branches followed such an exact pattern. They did not all grow exactly to the limit of the shape, but unless a branch was broken they rarely seemed to exceed the shape. The question arose as to whether the tree's external structure was a reflecting the structural or electromagnetic configuration of the cells. I decided that they probably did.

It may be that male trees have one crystal shape and female trees another shape, in the broad-leaved trees.

I decided to make a model to see what sort of crystal shape would reflect such an external pattern. The answer appeared to be - an arrangement of twenty tetrahedrons, which projected a hexagonal shape, no matter where you viewed it from. By slightly tilting the model, the outline would change in orientation. Either the hexagon had a point at the top and bottom, or flats at the top and bottom as in Figures 21(a) and 21(b).

This turned out to be an interesting model for several reasons. It kept reminding me of DNA whose molecular construction consists of hexagons and pentagons all inter linked. The structure also reminded me of two hands, whose fingers interlocked as in Figures 22(a) and 22(b). This may seem fanciful, but the human embryo's feet and hands are virtually interwoven like this and gradually separate during later months in the womb.

The other interesting aspect is a very personal one. When I was about twenty, I 'Invented' a rotary engine, which I estimated had a heat efficiency of about 65%. A Patent Agent told me it would work but it would cost too much to develop and manufacture. It main features were two rotary pistons which interlocked; and the interlocking pentagons of the crystal structure reminded me of my engine. I was naturally pleased that my simple design appeared to be the same as that used by the cells. However, the crystal structure, like my engine, is probably of no importance.

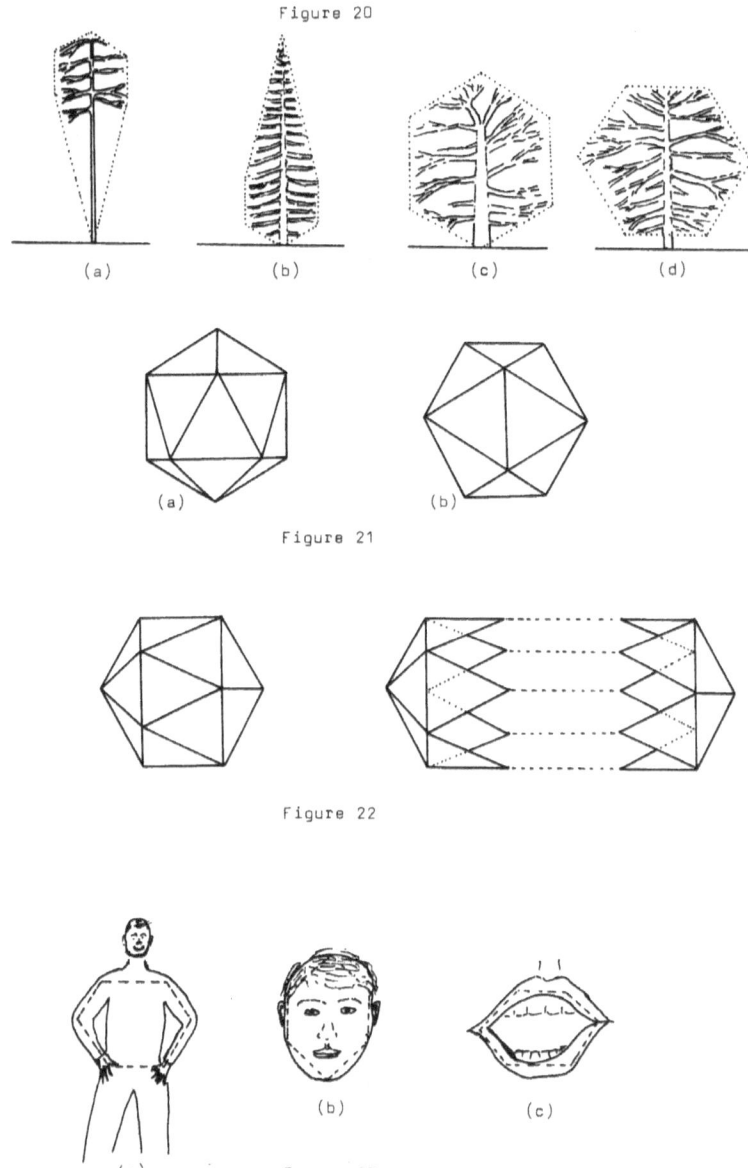

Figure 20

Figure 21

Figure 22

Fugure 23

Chapter 30

Crystal Humans?

After considering the trees, I went on to look at the structure of humans. Did we reflect simple crystal shapes too?

I stood in front of a mirror with my hands on my hips. A simple enough pose. I found myself looking at a hexagon, as in Figure 23(a). I looked at my face. At its simplest, allowing for rounding off, it too was a hexagon. I opened my mouth, and found to my surprise that it opened in the form of a hexagon. Thus we have Figures 23(b) & (c).

A further examination showed that allowing for roundness here and there, my chest, hips, feet and hands were all fairly simple crystal shapes. The palms of my hands were essentially pentagonal and I immediately thought of the interlocking fingers of Figure 22.

It is easy to see triangles made up out of your nose, mouth and eyes, but what about your ears? I realised that viewed from above, my nose and ears were growing on lines which intersected at 120 degrees in the centre of my skull. I wonder why this is? I didn't particularly like the idea of being crystalline - it seemed slightly robot-like. However, it does appear that we humans are much more like crystals, trees, insects, and other animals than we may prefer to believe. It seemed likely that the model shape in Figure 21 would apply to us as well as to trees. This seemed a bit fantastic.

I considered the tilting of the model and its effect on males and females. The main features of men and women are the same, so any differences in structural shape, as dictated by the cells, must be very subtle. In Chapter

23, I had decided that the male uni-cell actually entered the female uni-cell, and that the orientation of the male uni-cell would decide the sex of the child. The difference in orientation would have been very slight, but this could account for the tilting effect of the model.

I worked out the effect of the tilting, in human terms, and immediately ran into a slight oddity. The tilting would mean that a man's head would be at a different orientation to that of a woman. It would leave men looking upwards and women looking downwards because of a variation in the respective relationship of their eyes and ears. The men's eyes were slightly higher than their ears, while the women' eyes were slightly lower than their ears. If they were to converse on 'level terms, the men would have to have lowered their heads a little and the women raised their heads a little. This seemed absurd.

On the other hand, did this is explain why women seemed to have longer necks and more horizontal jaw lines than men?

Chapter 31

Binary Stars

This chapter is a rather strange one and given the context of this book -- that's saying something. I have included it because it seems to be important. It follows the thinking of Chapter 25 in a more ruthless, 'inhuman', explicit way, and I must say it has been rather unpleasant to write, in some ways.

The reader may prefer to skip this chapter and go on to the next one.

The Natural Historians, who follow the Darwinian theory of evolution by natural selection, appear to imagine that species simply develop by accident; and those 'accidents' best suited to their environment were those which survived and went on to form the basis of later evolution. In my view this is wrong.

As I stated in Chapter 25, I think evolution in living organisms simply parallels evolution in elements and compounds. In other words, the universe is a large evolving chemical and atomic reaction. As new elements are evolved and multiply, so the relative concentration of elements and compounds alters. This change in concentration causes the effects noticeable in the evolutionary time wave.

We human beings are part of this universal reaction. We are an integral part of the universe, and as such, we are simply derivatives of hydrogen reacting with other derivatives.

If it were not for nature, this universal reaction would be entirely organised and completely predictable. The 'life' structures exist where the temperature on the time wave promotes a reaction in semi-solid form. 'Living things' are seen as semi-solid while non-living things are either solid, liquid or gas.

We humans call things 'artificial' when referring to man-made objects. However, we tend to use the word to distinguish such objects from natural objects. Things are considered to be either natural or artificial.

In reality - all things are natural.

A telephone is a natural growth produced by an animal organism (humans). Jet planes, skyscrapers, juke boxes, and pollution are all natural. The stars make new elements. The planets make new compounds and new elements. Humans also make new elements, but since they are part of planets, and part of the universe, this is hardly surprising. Trees produce oxygen. Humans produce plutonium.

So nuclear reactors are a natural phenomenon as are their products: nuclear weapons, electricity, and new elements or compounds. If trees produce oxygen, and this caused 'mutations' of trees namely: insects, animals and humans; it can be seen that 'our' products are simply part of an evolutionary process which will produce human 'mutants' in the future.

It may be comforting to assume that future mutants will need the 'old' human species to produce the new elements and compounds upon which the survival of the future mutants will depend. Such human 'mutants' evolved from new nuclear elements should perhaps, therefore, be seen as an advance along the evolutionary time wave.

The question arises as to how the universe can ensure that we can survive the products of our nuclear evolution. It is my guess that it cannot. The joker in the pack is nature. There may be a sudden, unexpected accident.

However, the universe of Hydrogen is very organised and builds in a series of checks and balances to prevent cataclysmic accidents. In this context, this book may be seen as a balance to our nuclear power. The knowledge it may bring in terms of social self-control may prevent future large-scale wars and thus save us from self-extinction.

If the human race was completely wiped out, this would be a setback for Hydrogen, as it would prevent the continued evolution of nuclear products.

However, the evolutionary time wave would continue, and a 'new' human race would be formed. Such a race might not be exactly like the present one, -- indeed it would probably have to be different to avoid a second mistake.

In my view, such a universal setback due to self-extinction would be unlikely, especially as there are so many other forms of displaced energy which could do the job much more effectively and simply. A very minor alteration in the outer wave structure of the Earth could do far more damage than any nuclear war.

In my opinion, the Earth is expanding and not contracting. I believe it has been expanding since its conception. The expansion is caused by the continual aggregation of elements from space, including photons etc. The Earth grows like a cell, from the inside. The solid part is probably fairly thin, although thicker than most people think.

It is a sobering thought that just as cells can split when their outer compounds get on a collision course with their faster or slower moving centres; so the Earth could split in exactly the same way and for the same reason. It seems likely that this is how binary stars are formed. I do not think this is how the Earth was formed.

Chapter 32

Universal Measurement

In this book, I have been much concerned with the structure and simple patterns which appear to point the way to an evolution in knowledge. The ability to see these patterns and structures does not seem to belong to everybody. The ability seems to depend partly upon a need for knowledge, and as such would only belong to knowledge seekers. It also seems to depend upon the ability to observe the simple, basic patterns in a way which is free from the confusion of apparent reality.

A combination of knowledge seeking intelligence and simple stupidity seems to produce the paradox necessary to understand something of the Universe.

I have often asked myself stupid questions in order to get paradoxically sensible answers. However, my own intelligence has often lead me to clever but incorrect answers. From all this experience, I have concluded that my limited ability to observe simple things is because I am a very average individual. Average in the sense of being a combination of opposites, rather than uniform.

This averageness enables me to see both sides of an argument, and to be equivocal and contrary at the same time. The observation of patterns and outlines, in an intellectual sense, can only be achieved if the medium of observation is related to the observer. It is important, therefore, that we use measurements and scales based on human measurements if we are to perceive with human senses.

In relative terms this will allow us to attune to the universal scales and measurements common to all things. If we accept that most things in the universe are derived from hydrogen, then surely we must accept that all such derivatives must be related in terms of dimensions. Both linear and time dimensions should be seen to be interchangeable and to correspond with human dimensions. Galaxies, solar systems, planets, moons, elements, compounds, organisms, life - humans, all effectively the same material in different forms; all related, all to the same common scale. This is universal measurement.

In my view, measurements and scales evolve with the evolutionary time wave and those which are closest to the universal measurements will survive. Thus many ancient scales have been discarded. In this century, England has seen many linear scales dropped from common use. We are turning from the linear scale based on the foot to that based on the metre.

It may be thought that the transference from a human 'foot' measurement to a platinum bar measurement goes against my conception of survival based on human measurements. This is partly so. However, the metre is closer to the universal scales that the foot. Thus it will survive when the foot is discarded. How is this?

The 'foot' is being discarded because it is a distortion. Like the 'hand' (used for measuring the heights of horses), and the cubit, the foot is to disappear.

Human beings have four feet, but the English linear measurement was taken from the rear feet only. An average of both front and rear feet produces a shorter 'foot' a ten inch foot. (The length of a hand is about 8" and a foot 12"; so 8"+12"= 20"; 20" divided by 2 equals 10"). Four 'ten inch' feet equals about forty inches which is virtually one metre. Hence the metre is a more 'human' measurement than an English 'foot'.

Complete decimalisation will bring more distortions. We must try to conform to human based measurements in time and linear scales. A heartbeat is probably closer to universal measurement than one second. All measurements should fit a common pattern or ratio. There will be paradoxes here, as in all things.

Chapter 33

Language of the Universe

While I was thinking of the common scales and measurements of the universe, I considered the survival of language. Why was it that the English language was so popular? Why had it become the major language of international business and science.

We all know it is by no means perfect language -- the English have more difficulty with it than most. Nevertheless, in an age of mass communication and social evolution the idiomatic languages seem to be more popular. In my view, this is because idiomatic languages, such as English, are closest to the language of the 'universe'.

The 'language of the universe' is the mechanism of transference in space and time, conducted by waves. All languages are forms of modulated wave patterns with an identity which is understood by the organisms which produce them. It is my hypothesis that language is sound produced by organisms as a result of their reaction processes. Each sound represents an electromagnetic or structural concept which the cells can identify.

The sounds we humans identify, via our ears, are wave patterns produced by changes in the energy structure of different parts of the universe. We do not 'hear' all the sounds because our ears are only sensitive to certain narrow wave bands.

However, the sounds we 'hear' are only a part of the language understood by the cells. Other wave patterns exist in elements and chemical compounds and these form the basic means of identification for our cells. Language is important because it can enable us to understand the reaction processes of our cells. In my view, our human languages all represent both phonetically and in written form, the electromagnetic and structural processes used by the cells.

The basics are bound to be very simple. An infinite variation of process stages is much more likely than a large number of separate non-interchangeable reactions. This is like comparing the simple English alphabet with the enormous list of complete entity Chinese characters. The cells are much more likely to use various combinations of a small number of interchangeable processes.

The acceptance of English rather than Chinese is a reflection of the close similarity of the mechanics of English, based on a small number of interchangeable letters, to the cellular mechanics of cells. Thus the world community is gradually adopting the same simple and effective techniques used by the cells which make up the individuals of that community.

An evolving species needs an evolving language. If a species cannot evolve its language, it will fail to identify newly evolved compounds and conditions. Therefore any evolving human species will require an idiomatic language rather than a fixed, structured language.

If I am right in thinking that all languages of the same species have a common base, it should be possible to find out which are the sounds and written symbols common to all. By comparing the actions these sounds represent with the actions of the cells, it should be possible to isolate the electromagnetic meaning of each individual letter. If words can truly represent a string of electromagnetic or structural processes, by virtue of the order of their letters, or the wave pattern of their sounds, the meaning of many unsolved conditions could be understood. Nearly every disease has a common name in all languages.

By analysing the way we speak, we may be able to compare the speech mechanism involved with the sounds produced. It seems likely that the tongue acts as a switch, and together with the mouth, lips, jaws and teeth produce the sounds which make up each individual letter.

We have already seen in Chapter 10 on time projections, that the Arab Numerals may have been based on such projections. The English letters 'O', V', 'R', 'S' & 'E' all seem to figure in these numerals. It tends to reinforce the view that letter symbols are probably based, perhaps unwittingly, on variations of energy projections. This makes me wonder whether the basic and often repeated word -- "love" can be represented in simple time conceptions.

It appears that love is a contradictory mixture of compatibility in electromagnetic structure and synchronisation of wavelengths. If the mixture doesn't manage to achieve both forms of compatibility then 'love' will' be short lived.

So, what of the letter symbols. 'E' represents an outward or inward motion of a spiral. 'V' seems to represent the complete variability of time paths for a moving particle which projects a sphere. 'O' is the time projection of a particle moving in a straight line. But what the 'ell can 'L' be (if you will excuse my French). Can we see something in the word 'light'? It seems rather visual -"look". Definitely something to do with the senses - "listen".

Let's be fanciful. 'L' to receive waves; 'O' to move in a straight line; 'V' to synchronise with all waves and 'E' to interlock with an opposite 'E'.

Lastly, I would like to consider the role of babies and language. Babies are international. When they are born, they all speak (or cry) the same language. I think this is the closest we humans ever get to the language of the universe, perhaps because babies are closer to the universe than most.

The baby's first cry is "Arhh", followed by an intake of breath - "Huh". The well-known sobbing of all humans. "Arhh-huh", "Arhh-huh", - a very basic sound. Next the baby may dribble. Some saliva will probably go down its windpipe. The result its first cough, - "kehh". A multiple coughing action produces 'Kehh - huh", "Kehh-huh", as it coughs, breathes in and coughs again. We symbolise the "Huh" sound with an H. 'H' is also the atomic symbol for hydrogen. Is this accidental? The "Huh" sound is the basic sound of breathing through the mouth - of panting.

By considering the sounds of a baby just after birth, we see that we have an "Arhh" for crying and a "Kehh" for coughing. 'Arhh-Kehh'. How similar to A.C. The letters A.C. represent alternating current in electromagnetic terms, perhaps in language too. The electromagnetic terms AC/DC are very basic and represent either a moving stream of electrons or an oscillating stream.

It may be that precisely the same electromagnetic conditions control a baby's coughing, crying, breathing and talking. Perhaps we should stop looking for meanings in 'adult' sounds until we have mastered the basics of 'baby-talk'. The simple and basic phonetics of children are the base upon which all adults build. Thus our symbolism is likely to be nothing more than a vast variation on their basic theme. If we can crack their code and their language, the rest should be child's play. Joke.

Chapter 34

Structure and Society

This book started off with a career prediction hypothesis and ended up with a hypothesis on the meaning of language. Throughout the book runs a common theme which relates the structure of the universe to the continuous evolution we call life.

It is possible that too much has been made of the structural influences and too little of the paradoxes of nature with its infinite variability and random unpredictability. However, if we are made of hydrogen derivatives, it is surely inevitable that we should write according to the logic and predictability of hydrogen.

My last chapter concerns the apparent effect of structures on the behavioural actions of our society. The early chapters clearly indicated that our individualistic societies tended to become dominated by the public response seekers. In this, they repeat the domination typical of that found in the communal societies.

Only when the environmental conditions have produced a lot of knowledge seekers is the governmental position altered. It appears that only knowledge seekers can achieve the social and economic advances in government which are necessary to maintain the civilisation which is unique to individualistic societies.

Public response seekers are public performers -- and as such we may expect 'actors' in various guises, in leading positions within our society. Thus we have Advocates and Judges; Lecturers and Professors; Priests and Popes;

Sergeants and Generals; Shop Stewards and General Secretaries; Managers and Tycoons; Councillors and Cabinet Ministers. All are invariably public response seekers who will wish to claim the centre of the public 'stage' whenever possible.

They will continually seek publicity - good or bad, and will tend to be conservative, sticking to the routines which bring them the greatest applause or attention.

The 'time' aspect of the universe appears to relate directly to the performance. Babies and the very old are closest to the universe, while the middle-aged are furthest away. I would expect the young and the old to understand the universe best, -- and therefore understand a society produced by the universe. Those in middle-age would tend to understand society the least.

A quick check through history seems to confirm that leaders appointed in their middle-age, i.e. 40-50 years old, have proved disasters for their societies. The best years seem to be around 25% or 75% of a politician's life span.

Perhaps the reasons for this are obvious. Both young and old are close enough to the universe to understand society and act in a sensible manner. The young have courage, vision, optimism and energy, with which they can propel a society through rapid technical and social advances. The old have the patience and experience necessary to ensure a consistent and mature approach to government.

The middle-aged leaders have neither courage, vision, patience nor experience. As public performers, however, they are poised and polished. It is no accident that such leaders engage in continual brinkmanship. This is necessary to ensure that the spotlight of public attention remains firmly fixed upon them. As history shows, the risk attendant on such brinkmanship is very high, since it often results in wars and internal strife.

In government, soldiers and priests seem to rule communal societies, while politicians rule individualistic societies. The question arises as to how they manage to rule. In an Ape society, the high ranking males rule by intimidation. The leaders of human societies seem to adopt the same basic methods. The soldiers rule by means of social terror. The priests rule by means of individual terror. The politicians rule by balancing the social and individual terrorists in society.

The soldiers 'perform' in battle, and to achieve this they must have enemies, whether real or imagined. Skilful military performers will maintain an atmosphere of social terror by means of propaganda designed to produce self-intimidation. This produces a reflex action of social aggressiveness which supports the soldiers in any actions they may undertake. Soldiers always carry with them the mark of melodrama. Everything is slightly exaggerated. Fear and terror, explosive action, gaudy uniforms, parades, medals and ribbons - actors all.

The priests are slightly different. A bit more subtle. They are the masters of individual terror. Their followers are invariably those who are timid, frightened, respectful and lacking in social courage. The priests often claim to be the agents of the fear producing phenomena. They maintain their position by encouraging acts which will increase the individual's mental discomfort. Invariably they talk in terms of an individual's guilt and perversion. They set impossible standards for their followers, and reap the harvest of failure that this ensures. A steady diet of failure produces the despair, despondency and fear of the future which is the priest's main weapon of terror.

The politicians are the law makers. They maintain a balance between the social terror induced by the soldiers and the individual terror induced by the priests. Too much of either could bring disaster to the individualistic society ruled by the politicians.

Social terror brings external wars, while individual terror brings internal strife and civil wars. Both forms of terror bring disaster to an individualistic society, although the memories produced by the strife they cause are the food of life for both the soldiers and the priests - breeding the well-known 'vendetta' symptoms common to most communal societies.

Perhaps we may say that individualistic societies are led by law making civilians who are public knowledge seekers when the society is advancing and public response seekers when the society is declining. Communal societies are led by soldiers when they are advancing and by priests when they are declining.

The structure of the human cells also seems to affect society in terms of the individual's reaction to their environment. The layout of the social environment can be expected to have predictable effects on individuals if it is not compatible with the individual's cell structures.

In this respect there do not appear to be any grids or crosses in the cell structures. Angles appear to be anything but 90 degrees. Because of this I would have expected cubic or grid-like patterns to have an unpleasant effect on society. The structure of humans does not appear to accept right angled concepts. We know that motorists prefer roundabouts to crossroads, and that crime is invariably higher in grid pattern communities. Does this bear out my prediction? Maybe.

If I am right about the lack of right angled concepts, then I would expect the cross of Christianity to be unpleasant also. There has, unfortunately, been violence and perversion right from the inception of Christianity; but this may be simply due to the type of priests it attracts.

I am comforted in this respect by the pictures of Christ on the cross which show the man to be slumping down into a 'Y' shape. This 'Y' shape seems to occur a lot in human structures, and it certainly occurred in my cell model. Perhaps the Christians should concentrate more on the man, and his teachings and less on the guilt associations of the cross.

The Jewish star of David is hexagonal, and as such more acceptable within human structures. The Islamic pentagon is also acceptable. However, the model suggested an inter linking of pentagon and hexagon rather than separation. Perhaps these religions can expect trouble due to their isolated approaches.

It is interesting to note the Astrologers' twelve sided Zodiac which fits the concept of the two hexagons, one male and one female, both making up the human entity. This may not have been the reason for its design, but it may explain the reason for astrology's survival and persistence.

The cell model actually contained 'X's & 'Y's, hexagons and pentagons. Perhaps all religions got somewhere near the answer and only need to link up to achieve complete identification with the structure of the universe in human terms. I wonder if the religious knowledge seekers can ever detach themselves from the 'priestly' public response seekers for long enough to achieve such unity. Maybe.

The structure of evolution is 'life' in human terms. Our society lives by converting the raw materials of the Earth into the goods and services society requires. The more efficient the conversion, the richer the society. The smaller the society, the greater the riches per individual.

Individuals tend towards the average; societies tend towards the extreme.

Individualistic societies can bring technical and social advances which benefit all, but they have proved difficult to maintain. Communal societies bring no advances but are very stable with their emphasis on tradition.

Growth in cellular terms can be towards fatness, thinness or equilibrium. The two former conditions bring a rapid decline in individual health, and an early death. There appear to be parallels in society also.

An individualistic society 'lives' by the exchange of goods and services between its individual members. There appear to be three types of exchange. Fair exchange, where individuals are fair to themselves and others. This leads to equilibrium. Unfair exchange, where individuals are unfair to others. And Unfair exchange where individuals are unfair to themselves. These last two lead to rapid declines and death. Declines and death, that is, of the individualistic society.

Unfair exchange leads to a communal society and thus halts civilisation. Only fair exchange maintains the equilibrium required by an individualistic society if it is to continue to advance in technical, social and economic terms.

It is interesting to note that both capitalism and communism lead to the same end. Both lead to a communal society and a halt in civilisation. Capitalism encourages individuals to be unfair to others, and communism encourages individuals to be unfair to themselves. Under capitalism, one individual ends up owning everything. Under communism, all individuals end up owning nothing. The results are the same.

It is surely one of nature's paradoxes that the systems end up producing the opposite results of their founder's intentions. Karl Marx envisaged a society of individuals, whose personal sense of fairness to themselves and others would make government redundant. The various prophets of capitalism seem to have had similar visions. Paradox upon paradox.

The final paradox is surely the 'growth point' of civilisation, and the individualistic societies which lead to it. It is not the strong rulers, nor the meek and weak who produce such societies. It is the tough, courageous outsiders, whose independence leads to knowledge seeking and subsequent civilisation. If a society is to maintain its advance, as the Roman civilisation did, it must continually look to such independent outsiders. The centre of society can be expected to be dominated by public performers seeking continual publicity, and courting the disasters which bring the decline of the individualistic societies.

If individualistic enterprise is replaced with the corporate enterprise of the communal societies, then the transition to the latter societies will be greatly hastened. In the past no one has ever understood the mechanism of such a transition. It will be interesting to see the results of such knowledge.

The evolutionary time wave continues. The challenge of life remains. In the past the communal society has invariably brought wars and internal strife which result from its soldier and priest leadership. Can we afford such wars and strife in the future? Perhaps we can.

The individualistic society has been very transient and difficult to maintain in the past. It needs the leadership of knowledge seekers, as well as public response seekers, to maintain the balance which produces peace. Can we afford peace in the future? Perhaps we cannot.

Here we have the paradox. Life and growth need a fair exchange which allows evolution without extinction. Other species rely totally on the environment to maintain that balance. Are we any different? Maybe.

J.D. Waters

This book contains many hypotheses. None have been empirically researched or tested. All may be completely wrong. I leave the readers to judge from their own experiences the value of these ideas.

They may be complete and utter nonsense, or they may be part of the language of the universe................ maybe.........

Appendix A
Career Prediction

1. Individual Response Seekers (IRS's)

    Requirements   Audiences of one; a degree of privacy; opportunities to see many individuals in series, every day.

    Careers   Accountants, Solicitors, Surveyors, Administrators, Bankers, Insurers, Clerks, Secretaries, Telephonists, Cashiers, Receptionists, Waiters, Representatives, Sales Assistants, Typists.

2. Public Response Seekers (PRS's).

    Requirements   Audiences of many; a need for 'self' or work to be publicly seen.

    Careers   Architects, Builders, Engineers, Actors, Entertainers.

                            Models, Professional Sportsmen, Barristers, Lecturers. Politicians, Priests, Artists, Authors, Designers. Journalists, Chefs, Gardeners, Stock Breeders. Surgeons. Craftsmen, Decorators, Production Workers. Armed Forces, Drivers, Pilots, Police.

3. Individual Knowledge Seekers (IKS's).

Requirements	Needs a learning situation which will improve understanding of 'self' or others; privacy and one-to-one situations.
Careers	Dentists, Doctors, Psychologists, Nurses.
	Probation Officers, Social Workers, Careers Officers.
	Personnel Officers, Instructors, Teachers.
	Friars, Monks, Nuns.

5. Public Knowledge Seekers (PKS's).

Requirements	Needs a learning situation with opportunities to improve the environment or human society in a general way; prefers privacy; will 'lecture' rather than teach; needs to have work publicly seen, eventually.
Careers	Natural and Social Scientists, Economists, Historians.
	Philosophers, Research Workers and Technologists.
	Scientific Officers and Technicians.

Appendix B

Human 'Archetypes'

<u>'A' Types</u>	Negative Top. Positive Bottom. Mostly Female.
Food & Drink	Negative tops, so: preference for high energy food and drink; Cakes, Spicy foods, Curries, Fried food, puddings, cream, sweets, gin, vermouth, sherry, white wine. Hot tea or coffee with milk and sugar.
Careers	Prefers sedentary occupations
Marriage	Prefers 'V' & 'H' types.
General	Men may compensate for shape by being more polite, gentlemanly, and tidy; they may grow beards to emphasise masculinity.
	Women may balance their negative top with 'positive' jewellery: tiaras, necklaces, earrings.
<u>'V' Types</u>	Positive Top. Negative Bottom. Mostly male.
Food & Drink	Positive tops, so: low Energy foods, savoury rather than sweet; boiled, braised or grilled; Cool liquid drinks: Beer, Chilled wine, Red wine, Whisky 'on the rocks', Milk, Lemonade.
Careers	Prefers active occupations.
Marriage	Prefers 'A' & 'H' types.
General	Women may compensate for shape by being aggressively

	Sexual with low necklines, high hem lines, long hair. They may also wear very high heeled shoes to 'suggest' stability.
	Men may wear large, country type shoes.
'H' Types	Positive/Negative Top & Bottom. Even Males and Females. Usually Attractive.
Food & Drink	No special preferences.
Careers	Part sedentary, part active. May travel a lot.
Marriage	All types. Children follow partner: daughters if 'A' types, sons if 'V' types.
General	Even tempered, equivocal, contrary, and indecisive, tend to be lazy.

Epilogue

Author's Time wave

If there was any doubt about my poor intelligence, it was duly confirmed after I had taken my mid-term examinations. The Examination Board of the South Bank Polytechnic decided unanimously that I had not reached the academic or professional standards required and I was told to discontinue the Course.

As a result the London Borough of Brent, who employed me as a Trainee and had seconded me to the South Bank Course, asked me to resign from their Careers Service. Such is the price of stupidity.

Thus I ride my own evolutionary time wave. Down today. Up tomorrow… Maybe.

References

Bibliography

Bronowski, Jacob. 'The Ascent of Man'

British Broadcasting Corporation 1973

Goodhall, Jane van Lawick. 'In the Shadow of Man'

William Collins 1971

www.ingramcontent.com/pod-product-compliance
Lightning Source LLC
Chambersburg PA
CBHW030814180526
45163CB00003B/1284